R.E.A.L.I.T.Y.

For all seekers of the truth

R.E.A.L.I.T.Y.

Resolution of Epistemological Analytical and Logical Issues in the Theories of Yesterday

Peter Jewsbury

Email: peter.jewsbury@gmail.com

2006

Published by *A*libro *A*ustralis *S*cientifica, Melbourne, Australia 2006.

First edition.

First published in Australia in 2006 by *A*libro *A*ustralis *S*cientifica.

This book is available for internet purchase in different bindings at web site www.lulu.com.
Shopfront web address: http://stores.lulu.com/store.php?fAcctID=644434.

ID: 453454/ 506160 (paperback)
ID: 555304 (hardback with dust jacket)

Description: xxxvii, 280 p. 3 ill., 29 tbl. ; 25 cm

LC Classification: QC20
Dewey Class No.: 530, 501, 110, 113, 121
Physics and Astronomy Classification Scheme: 01.70.+W , 03.65.Ta

Cataloguing:
Fields: Physics, physical laws, foundations of knowledge, philosophy of science.
Search keywords: fundamental, nature, eventual reality, ultimate question, universe, reality.

Jewsbury, P.
R.E.A.L.I.T.Y. : resolution of epistemological analytical
and logical issues in the theories of yesterday.

Bibliography.
Includes index.
ISBN 9780980315103 (hbk.).

ISBN 0 9803151 0 7 (hbk.).

ISBN 9780980315110 (pbk.).

ISBN 0 9803151 1 5 (pbk.).

1. Science - Philosophy. I. Title.

501

ISBN-13: 978-0-9803151-0-3

ISBN-10: 0-9803151-0-7

Prelude

This document presents a radically new overarching paradigm for science with an accompanying new theoretical description of all reality. The new hypothesis that delivers these descriptions is called the hypothesis of eventual reality. Despite the foundation of the hypothesized reality being so far removed from our experienced everyday reality, the new theory is remarkably easy to visualize and easy to comprehend. Indeed eventual reality is highly unusual for a theory; the theory claims (in principle) to be a no-parameter theory, it claims to provide a theoretical description of everything without exception; furthermore, the theory is derivable from a very small set of self-evident tenets ie the theory is completely self-justifying. These features set the theory apart from the historically earlier *scientific* grand schemata eg classical Newtonian reality, quantum reality and relativistic reality.

Indeed, it is argued that all of the fields of science are derivable from the tenets of eventual reality. Scientific questions about the dimensionality of space, the flow of time, the origin of electrical charge, the number of elementary-particle generations, etc. can all be convincingly answered by a simple form of the theory. Eventual reality also answers philosophical questions about universal existence, the laws of nature, consciousness, etc. Furthermore, it provides a sound foundation for many traditionally distinct disciplines external to the sciences including mathematics and logic.

The introductory (approximate) forms of eventual reality that are presented here are intentionally simple and hence these introductory forms have several aspects requiring more detailed confirmation. These simple forms are used to verify that the picture of reality provided by the hypothesis of eventual reality, delivers known scientific principles. The first simple form was achieved by requiring that space be flat (contrary to evidence provided by general relativity). Matter, space and mind are clearly explained as derivative concepts. The second simple form shows that (in the classical limit) when spatial curvature is permitted we recover a form of general relativity with cosmological repulsion.

The work reported in this document is divided into three parts; each part has different aims. Part 1 describes the methodology that was adopted to devise a new hypothesis upon which (physical) science could be based. A set of believable values about the requirements for a new theory, is presented. Then these values are used to engineer a guidance framework for generating an alternate worldview. These steps are followed and a new picture of reality is sketched. In part 2, this picture is re-presented as a set of founding tenets that generate the new worldview. These tenets are turned into a comprehensive theory of reality and several implications of theory are investigated. Then the founding tenets are show to be self-evident. Now, in part 2 of this work, the quantitative implications are made readily investigable by approximating space to be Euclidean flat. In part 3, the flat-space simplification is lifted but the restriction is replaced by another approximation that represents matter classically. In this way, part 3 shows that the theory of eventual reality also subsumes the theory of general relativity. Indeed one common simple principle, the first principle of the universe, subsumes both theories.

The author's expectations, at the commencement of this work, have not been realized. The author anticipated that science was an unending downward quest – the final ever expanding frontier for discovery. He expected that the new reality (if uncovered) would be just the next deeper step in science's progress. Furthermore the author suspected that as we progressed further away from the classical world, the key concepts would become less familiar and more mysterious. Put simply, if quantum mechanics and general relativity are difficult to explain to the novice how much greater should be the difficulty that we might expect to encounter when explaining a more fundamental theory (even further from our experiences) that is to replace both quantum mechanics and general relativity.

However, this work has achieved unanticipated outcomes. Science merges into natural philosophy but the reductionist trend of unfolding deeper and more fundamental pictures of reality, is found to stop suddenly. This sequence of pictures is referred to as the multi-storey edifice of reality. In retrospect, it is easy to justify why this multi-storey hierarchy should have a foundation. However, sometimes one needs to be confronted by a conclusive result before the expectation of that consequence becomes undeniable. Furthermore, the uncovered ultimate reality is elementary and hence non-mysterious. Thus whilst our language is rather lacking in appropriate words to describe the foundation storey of reality, the pictures that are developed of this ultimate reality and of this hierarchy of storeys built upon this foundation, are quite vivid. In addition our intimate association with and participation in the aethereal proceedings of the universe can even justify the reason why the universe is so clearly explainable.

Now, the naming of this new theory occurred quite early in the development of this work and the name seems to correctly imply an expectation of a foundation to reality. However, in part, the name 'eventual reality' was conjured to ridicule the pretentious idea that such a reality would exist! Post partum, the name of the newly born theory, should be taken to mean that the theory applies to every storey of reality. This multi-storey applicability is another unusual feature of the hypothesis of eventual reality. Previously individual major theories have spanned no more than two storeys (eg statistical mechanics derives the relations of thermodynamic from a molecular viewpoint). However, eventual reality describes all storeys. Hence in this work, at least five storeys of reality are described in some detail.

The early roots of this work lie in an old undergraduate lecture course entitled *An Expedition to the Foundations of Physics*. This course was written and delivered by the author, many years ago, as a postdoctoral researcher. However, in 1997 he began to recollect and reconsider the questions that were raised in this course. By 1998 the author had decided upon an approach that has the potential to shed further light upon the structure of reality. Hence, purely as an intellectual exercise, he would often spend an hour or two at weekends casually plumbing some small facet of reality. It was never clear whether this work would achieve anything useful or how long the work would take to reach a conclusion. There were always too many open questions for any serious thought of publication. As events intervened, the work would be put on hold for several months at a time. However, towards the end of 2004 a picture of the theory and how the theory could be presented was gradually emerging. The author decided to retire in January 2005 and as a hobby has continued to develop, formalize and document the theory. This document is the outcome of these developments.

Finally let us summarize the aims of this write up. This document presents an introduction to the hypothesis of eventual reality. It is written for a general audience with knowledge of the physical sciences. Since it is an introduction, only the simplest forms of results are derived. This introductory work is divided into three distinct parts. The arrangement of topics replicates the order in which the work was undertaken by the author. In part 1 of this work, we are primarily concerned with developing aids to assist with identifying the details of a lower unknown storey (of the edifice of reality) from higher known storeys; whereas, in parts 2 and 3, we are concerned with building, exploring and developing the precepts that allow us to climb from the lower storeys to the higher storeys. In these parts we reveal known science and confirm the utility of the new hypothesis.

R.E.A.L.I.T.Y.

Resolution of **E**pistemological **A**nalytical and **L**ogical **I**ssues in the **T**heories of **Y**esterday

PART 1:
EVOLUTION AS **V**EHICLE FOR **E**LUCIDATING **N**OVEL **T**HEORETICAL **S**YNTHESES

PART 2:
AN **E**XPOSITION ON THE **T**HEORETICAL **H**YPOTHESIS OF **E**VENTUAL **R**EALITY

PART 3:
SUPPLEMENTING THE **P**ARADIGM WITH **A**FFINE **C**URVATURE **E**XPRESSIONS

CONTENTS

PART 1:
EVOLUTION AS VEHICLE FOR ELUCIDATING NOVEL THEORETICAL SYNTHESES

PART 2:
AN EXPOSITION ON THE THEORETICAL HYPOTHESIS OF EVENTUAL REALITY

PART 3:
SUPPLEMENTING THE PARADIGM WITH AFFINE CURVATURE EXPRESSIONS

SYNOPSES

Document
This work introduces the hypothesis of eventual reality. Eventual reality is promulgated as a 1) self-evident, self-authenticating theory of reality; 2) scientifically based theory extending beyond science into all areas of rational knowledge; 3) useful parameter-free fundamental theory; 4) theory requiring the existence of both mind and matter; 5) theory incorporating both quantum mechanics and general relativity. It is shown that the much-vaunted theory of everything quantitative is simply Fermat's principle.

Prelude
A new scientific worldview of reality is given in this document. Simple approximations are made to derive known forms of science. The presented vivid picture of reality merges science into natural philosophy. Part 1 describes how the hypothesis was developed. Part 2 and 3 show how the theory explains everything.

Preface
A table of word usage is given. For instance 'reality' is that which is required by science, 'existence' means membership of the set of real things.

PART 1: EVOLUTION AS VEHICLE FOR ELUCIDATING NOVEL THEORETICAL SYNTHESES

Prologue to part 1
Tackling the nature of reality head-on has not born fruit in the past. Perhaps an easier approach is to exploit the similarities in how science advances.

CHAPTER 1 : THE SEARCH FOR MEANING

1.1 Introduction

Major shifts in scientific worldviews exhibit a recognizable pattern. A map formed from this pattern would be a useful guide for further evolution of the worldview.

1.2 Guide map

Four meta-principles will help us construct a map to describe the accomplishment of a paradigm shift. Constructing the map is the first step of the map's journey.

1.2.1 History rewritten

Quantum mechanics was developed over a quarter of a century and had distinct contributions from many scientists. We can recategorize these contributions to make a rational story of how quantum mechanics was developed.

1.2.2 Copenhagen interpretation of quantum mechanics

The standard interpretation of quantum mechanics is minimalist, permitting only the least amount of attributed meaning to quantum mathematical expressions. Notional quantum reality can only be attached to those features which have been measured.

1.3 Dark clouds

The second step of the map is to identify problem areas where current theories inexplicably fail. We ask what foreign concept we can add to patch up the difficulties.

1.4 The master principle

The third step is to generalize the suggested foreign concept into an overarching principle with wide applicability.

1.5 Dualities

The fourth step is to identify what related but seemingly incompatible concepts are preventing us from achieving a new synthesis expressed in terms of more fundamental concepts.

1.6 The evolutionary environment

The fifth step is to examine what evolutionary scientific trends and ancillary theories can assist the development or promotion of the new theory.

1.7 Finding the starters block

The sixth step is to assemble these answers, which are pieces of a puzzle, around what is truly real and thus to provide a sketch (not yet a formalism) of the new paradigm.

1.8 Aiming low

The seventh step is to be cautious when constructing the new formalism by making some familiar approximations.

1.9 Evolutionary survival

The eighth step is to share the conclusions of this analysis and, if appropriate, promote the theory's value.

CHAPTER 2 : THE TROUBLE WITH QUANTUM MECHANICS

2.1 Introduction

There is clear evidence that quantum mechanics is not the final scientific theory. For instance, quantum mechanics has limitations, the theory includes *ad hoc* rules and it is missing many explanations.

2.2 The measurement dilemma

Measurement is restricted to otherwise isolate regions of the universe. Quantum mechanics has two incompatible parts, one describing the propagation of the state and the other part how to calculate the statistics of a measurement. Denying the reality of the wavefunction seems unreasonable. Yet its reality is contrary to relativity.

2.3 Coordinate representation of nature

Quantum states are partially represented using mathematical-complex algebra. We should fully apply this algebra if we accept what nature is telling us about the mathematics that is necessary to describe reality. A full application of this algebra would require both making coordinates of space mathematical-complex and adding a second spatial branch with an alternative time sense.

2.4 Time sense

The Copenhagen interpretation is time asymmetric and hence requires that the quantum state be unreal. However, this implication does not follow if we have branches with different time sense.

2.5 Space-time anti-discrimination

Quantum mechanics treats space and time as quite different variables. However, we may relativistically mix space and time if we allow measurement to scatter the state in both time directions.

2.6 Over-simplified model of free mechanics

We consider a model of a massless non-interacting particle obeying some form of statistical mechanics. We make all variables mathematical-complex. Interestingly this system is describable by a probability function resembling a plane wave.

2.7 Conclusion

The analysis suggests that quantum mechanics may be the random statistics of a properly constituted mathematical-complex analysis in which time and space are combined relativistically.

CHAPTER 3 : THE MASTER PRINCIPLE

3.1 Introduction

The master principle is to be a further generalization of the principle unearthed in step 2 of the map.

3.2 The pairing of observables

The interconnections of the formulations of classical and quantum mechanics are noted. There are associations that relate Fourier conjugate variables and quantum complementary variables. However, these associations leave arbitrariness in a scale factor.

3.3 Hyper-complex numbers

Scientific formalisms are based upon mathematical-complex numbers and vectors. Vectors may arise as components of quaternions, which are known to be a generalization of mathematical-complex numbers.

3.3.1 Quaternion essentials

Quaternion numbers describe a space of 3+1 dimensions. Non-analyticity mars scalar products of quaternions.

3.3.2 A generalized picture

The inclusion of quaternions in our expressions reveals a bigger picture seeking the involvement of more generalized algebrae. Symmetry requires that each system of algebra will introduce further branches to the generalized coordinate space.

3.4 Conclusion

Mathematical-real numbers poorly describe nature; nature is better depicted by the numbers of a less-restricted algebrae, these algebrae are probably those of mathematical-complex and quaternion numbers.

CHAPTER 4 : DUALITIES

4.1 Introduction

For all of recorded history, the clash of apparently incompatible concepts (eg permanence and change) has acted as an impediment to the advance of knowledge.

4.2 Modern dualities

Today there are eight potent scientific dualities. Many have been given some partial form of resolution by either quantum mechanics or general relativity. However, the two dualities, which are the most compelling of this set of dualities, are those that involve the 'qualitative and quantitative' and 'mind and matter'.

4.3 Qualitative and quantitative

There is a qualitative realm that is not amenable to quantitative analysis. This realm can be represented using a suitable language rather than mathematics.

4.4 Mind and matter

We have a deep intimate knowledge of mind. Yet conventionally attributing primacy to the concept of matter has not enabled us to understand the mind. To merge mind and matter we will need to forge a new union.

4.5 Conclusion

We shall adopt both the qualitative-and-quantitative duality and the mind-and-matter duality. These dualities have similarities. We shall use a linguistic approach to manufacture a unification of these disparate concepts.

CHAPTER 5 : CONTEXT

5.1 Introduction

Evolutionary fitness of a theory is judged by the theory's acceptance. Hence we should consider the environment of conventional ideas.

5.2 Blemishes in our understanding

Current scientific knowledge is limited in its understanding of science's basic concepts like space and matter.

5.2.1 Spatial dimensions

Extra spatial degrees of freedom are required to explain intrinsic quantum numbers. Some theories hypothesize unseen wrapped-up dimensions to provide this freedom.

5.2.2 What is matter?

We do not have a clear idea as to whether basic elements of matter are point particles, strings, branes or something else.

5.3 Competing explanations

Our restricted glimpse of the universe (from practically one time and one place) allows abundant cosmological ideas, often seemingly supported by observation, to be promoted. Some of these theories such as those involving cosmological inflation and multi-worlds may need abandoning. In the absence of more observational data, the devotees of these theories may not be easily persuaded.

5.4 Tidal currents

As new science displaces old science, the old science is generally seen as a macroscopic approximation synthesizing coherence from a less structured underpinning description. The older explanation is comprehended as the obvious conclusion of applying Fermat's principle to the corresponding underpinning reality.

5.5 Conclusion

We conclude that we are in need of a deeper and freer reality to underpin our concepts of space and matter. Furthermore, Fermat's principle should deliver this existing scientific reality from the more basic reality.

CHAPTER 6 : NEW VISION

6.1 Introduction

The results of the first five steps are summarized in a table. The emerging picture describes a reality built from qualitative options upon which a simple quantitative reality in unspecialized numbers is constructed by application of Fermat's principle.

6.2 Quale

We wish to describe the universe of fundamental qualitative options as an easily comprehended model. Disregarding all the trapping of a quantitative worldview we notice that what remains are referential pointers. These pointers are precisely what we need to fabricate a picture of the foundation of reality.

6.3 First principle

Fermat's principle is the first principle of the quantitative universe. It describes the crystallization of increasingly certain information.

6.3.1 Classical stationary weight

A sum of intensity contributions peaks near the maximum of the weight (statistical measure) function.

6.3.2 Classical stationary phase

A sum of phaser components will peak when many of the contributions add in-phase. Organizational symmetry and the prevailing boundary conditions determine where spatially the set of contributions will be predominantly coherent.

6.3.3 Eventual stationariness

At the deepest level the contributions have no imposed symmetry or weight. These aspects are generated as a consequence of Fermat's principle.

6.4 Conclusion

We have generated a picture for an improved theory of reality. The instructions of the map have not guaranteed that this theory is correct; however, the instructions have made it likely that the theory possesses a close fit to reality.

Epilogue to part 1

The picture of reality that we have generated is clearer than numerous other suggested pictures since our picture was constructed by subsuming scientific experience.

PART 2:
AN EXPOSITION ON THE THEORETICAL HYPOTHESIS OF EVENTUAL REALITY

Prologue to part 2

A full picture of the new theory is given. A qualitative network of pointers, known as the apeiron, progressively generates the disciplines of logic, mathematics, science and philosophy. This apeironic foundation generates space-time and matter in understandable forms. Arguing from self-authenticating tenets the hypothesis needs circular 'by inspection' logical arguments to get started.

CHAPTER 7 : INTRODUCTION TO EVENTUAL REALITY

Eventual reality is a theory explaining all storeys of the edifice of reality. It draws definite conclusions. The aim is to demonstrate a satisfactory comprehensive picture of reality. Thus we derive a simple form of quantum mechanics to demonstrate that the theory reduces to quantum mechanics and not to determine intricate details of the quantum formalism.

CHAPTER 8 : THE MOTIVATIONS FOR CHANGE

8.1 Why displace quantum mechanics?

Quantum mechanics can only be applied to isolated regions of the universe. Quantum measurement and the quantum equation of motion are inconsistent. Quantum mechanics does not provide a model of reality. The philosophy of quantum mechanics seems to be incomplete and is incompatible with general reality.

8.2 Why rewrite the paradigm?

Material monism does not give a complete description of the universe. We do not have a consistent paradigm of science. We have too many models containing unexplained parameters.

8.3 The origins of eventual reality

The hypothesis was developed inductively from the core structures of conventional physics. An indication of how this was accomplished in part 1 is briefly recalled. Investigating the hypothesis will show whether the arguments of part 1 are useful.

CHAPTER 9 : A DESCRIPTION OF EVENTUAL REALITY

Our standard vocabulary is ill-suited to describing a timeless spaceless apeiron.

9.1 The tenets of eventual reality

The four foundation tenets of the hypothesis describe the apeiron, the foundation storey of reality. These tenets are semi-familiar. They are given as: 1) the absence of restrictions, 2) the primal reality is qualitative, 3) reality is structured in a hierarchical manner and 4) the foundation elements are pointers.

9.2 The foundations of space

Fundamental reality is a spaceless and timeless network of pointers that are known as events; this network is called the apeiron. Foundation space, called coordinate space, is feature of the first elevated storey of the hierarchy of representations of reality. Several principles underpin how space is formed as the resultant coherent summation of interlinked jumbled sub-networks of events.

9.2.1 Invariance and symmetry

Representations have a non-contingent symmetry indicating operations that leave descriptions invariant.

9.2.2 Representational groups and symmetry

Groups provide an algebraic way of specifying a spatial symmetry.

9.2.3 Coherence and symmetry

The incoherent sub-networked contributions cancel leaving as dominant the coherent set of sub-networks identified by symmetry.

9.2.4 Numbers and symmetry

A quantitative characteristic is introduced via the counting of sets of events.

9.2.5 Stationariness and symmetry

Space will exhibit symmetry associated with the space being stationary with respect to all variations. We seek two independent symmetries so that the space may possess recognizable displacements and rotations. These symmetries are represented by groups with operators $+$ and $\times$.

9.3 Fundamental coordinate space

Order lurks in the turmoil of the network of pointers. A foundation continuum space is formed as an eight-dimensional and eight-branched space. It is discussed how these dimensions and branches are manifest in reality. The eight-fold symmetry factorizes into independent subspaces of four and two dimensions. There are only five independent directions in the space-time energy-momentum foundation coordinate space.

9.3.1 Objections to a complex quaternion space

This foundation coordinate space does not possess the structure of the space that we empirically observe. Nevertheless it will be shown in later sections that our empirical space is a natural derivative of this foundation space.

9.4 The content of space

The hierarchy of substructures throughout space define a field called the aether. In the absence of purpose, the random statistics of this field define matter. Our material reality represents the combined statistical order that is endemic across the branches of space.

CHAPTER 10 : QUANTUM MATTER

Eventual reality describes a more fundamental reality from which quantum mechanics is sourced.

10.1 Macroscopic averaging

The analyticity that we associate with physical reality is simply a consequence of all practical spatial resolutions. Quantum mechanics is a bridge between microscopic eventual reality and classical reality.

10.2 Number forms

In unison, the theory of collective coordinates and Fermat's principle generate the hierarchy of distinct physical pictures. A reciprocal space (Fourier) transformation requires the introduction of an extension algebra involving quasi-numbers. Quasi-numbers form a module over the complex numbers.

10.2.1 Quasi-number extension algebra

The rules of quasi-number algebra are explored. Products involving quasi-numbers introduce new number designators.

10.2.2 Grassmann algebra

The new number designators are the basis functions for Grassmann algebra.

10.3 Collective transformations

A collective transformation of the aether is made using quasi-numbers.

10.4 Local reality

Particles (both real and virtual) are identified as discrete algebraic components of the statistics of the full aethereal field.

10.5 Mechanics

The characteristic properties of particle fields are revealed. The field elements are divided into fermion fields and boson fields. The acceptable form for an entropy parameter is shown to involve fermion and boson particle fields and their first derivatives only.

10.6 Variational principle

Fermat's principle gives rise to the second law of thermodynamics. A form for an equation of motion for fermions is deduced. Allowed symmetries are imposed to simplify the equation to a form resembling the Dirac equation of quantum mechanics.

10.7 Bosons

Boson fields are constructed from higher powers of the dressed particle fields. Repeating this undressing, we observe that there process generates only one fermion but potentially many bosons.

10.8 Contravariance and covariance

Covariant and contravariant variables are introduced as mathematical conjugate variables.

10.9 The measurement side of quantum mechanics

In the hierarchy, quantum measurement sits between quantum principles and classical principles. Measurement is discussed from both a mechanistic and a procedural viewpoint.

10.9.1 The mechanisms of measurement

Measurement apparatus is classical and like an electrical short circuit fuses the quantum disturbances on the different branches.

10.9.2 The character of measurement

Using our new description, we will consider measurements on quantum (not classical) systems.

10.9.2.1 Amplification

Measurement registers and sets one component of a complex number; the other complementary component is severely distorted.

10.9.2.2 Branch blindness

The overwhelming definite classical structure of a measurement apparatus spans all branches. The interaction with a quantum system neatly joins the possibilities on the branches.

10.9.2.3 Certainty

The definiteness of a measurement result follows from the classical complexity of the register of measurement.

10.9.3 CPT symmetries

The 'shorting' of branches during a measurement produces a CPT invariant outcome.

10.10 Summarizing the interpretation of quantum mechanics

Eventual reality attributes a (storey specific) reality to the quantum state thereby converting a measure of order into genuine information. Observation produces a backward-in-time signal; the indefinite causal chain leads to indeterminacy.

10.10.1 The quintessential quantum questions

We end the summary by listing the answers provided by the worldview of eventual reality to eight fundamental questions of quantum mechanics including: Why does quantum mechanics relate to experiment via probabilities? Can we apply quantum mechanics to the entire universe? What is responsible for the decoherence that is involved when a quantum macroscopic superposition adopts a classical value?

CHAPTER 11 : COMPARISON WITH THE TANGIBLE UNIVERSE

This chapter employs estimation and some empirical facts to discuss the broad implications of the theory.

11.1 Momenta

A macroscopic connection between the momentum and velocity of matter is deduced. It is shown why there is no universal position from which to measure displacement.

11.2 Hubble expansion

In the absence of any cosmological and gravitational forces, the time-scale 'constant' H is identified with the Hubble constant.

11.3 Information structures

The contribution of information, to particle rest mass, is quantized. Units are time dependent.

11.3.1 Time-dependent units

Although time as displayed in coordinate space, is unrestricted (ranging from $-\infty$ to $+\infty$, observational time is determined as a power of coordinate time such that observational time is always positive.

11.3.2 Large-number hypothesis

Cosmological repulsion ensures that orbits are likely to be unaffected by changing fundamental constants.

11.4 Particle charge and colour

The structure of the aether determines the quantum-number symmetries of elementary particles.

11.4.1 Electrical charge

Charge is identified with the number of quaternion-vector branches spanned by a particle.

11.4.2 Hadron colour

Hadron colour is just a label identifying each vector-branch.

11.4.3 Generations

Particle generation is identified with the number of vector directions entwined in the particle's information content.

11.5 Mass terms

The non-observableness of quarks means that we have limited approximate handles on the masses of quarks. Information is available from collisions and from contributions to the mass of composite particles. Thus primarily we shall discuss lepton masses.

11.5.1 Self-energy interactions

Hand-waving estimates of self-energies are given.

11.5.2 Leptons

An argument is constructed to give approximate charged-lepton masses.

CHAPTER 12 : BENEATH QUANTUM REALITY

Eventual reality can tell us about what descriptions of an underlying immaterial reality can justify and explain the cogently material representation of quantum mechanics.

12.1 Launching point

We shall discuss how to approach the representation and description of a qualitative aether.

12.2 Representation of the qualitative

The aether is a multi-level field of language – a dialogue in the universal langue.

12.2.1 Lexeme flavours

By analogy with our table of elementary fermions, we may identify lexeme flavours.

12.2.1.1 Quaternion symmetry

This symmetry is aligned with Carl Jung's four psychological types.

12.2.2 Complex symmetry

This symmetry is aligned with grammar of verbs and nouns.

12.3 Logic

Two mathematical signs represent logical values. Eventual logic does not accept the classical logical law of bivalence.

12.4 Mathematics

Traditional mathematics is seeded by the countability of events. Although absent on the lowest level of reality, the origins of mathematics appear self-consistently from symmetry on the first elevated level.

12.5 Mind

Mind is a necessary consequence of a picture of the universe formed as language. Ideas and decisions become fixed by quantum measurement.

CHAPTER 13 : PHILOSOPHICAL PICTURES

A paradigm is part of every scientific theory. The paradigm of eventual reality will be described and justified as self-authenticating.

13.1 Fabricated storeys of reality

A multistorey edifice is used as an analogy for hierarchical pictures of reality. The building of the edifice is described.

13.1.1 The primordial constructs

The foundation of reality is a boundless apeiron formed from a network of events.

13.1.2 Coherent-storey picture

Macroscopic order is naturally a part of microscopic chaos. On the first coherent storey, this macroscopic depiction generates coordinate space.

13.1.3 Matter-storey picture

Matter and our hybrid space are identified by recognizing cooperative behaviours.

13.1.4 Higher storeys

Quantum measurement and classical mechanics occupy the next two higher storeys respectively.

13.2 Jus naturale

Each of the founding tenets of eventual reality is explained as being self-evident.

13.3 The philosophical discipline

Philosophical antinomies are a consequence of the differences between the views of common concepts from the vantage point of different storeys of reality.

13.4 The great questions

In turn, each of 10 great questions of reality is explained. The chosen questions are:-

Why is the universe here? Could there be nothing at all?

What came before the start of time? How did the universe spring into being?

What is consciousness? What is free will?

How can we explain the passage of time? Why do we only remember the past and not the future?

Do we have souls? Is there life after death?

How do we describe existence in conceptual terms? What is the essential nature of the universe?

Why should we accept that there is a reality beyond our mental perceptions? How much subjectivity do we unwittingly impose upon the view of reality?

Why do the parameters of the universe appear to be finely adjusted for life? Could there be universes with different laws?

How can there be change if the apeiron is timeless? How can there be individual things if the apeiron is spaceless?

How can we claim knowledge and understanding of an underlying reality not amenable to experimental confirmation? How can an ultimate self-justifying storey of reality possess any non-trivial content?

CHAPTER 14 : THE CASE FOR EVENTUAL REALITY

Eventual reality has a distinctive paradigm, explains known scientific principles and makes clear predictions.

14.1 Elegant paradigm

Eventual reality claims to be self-authenticating. Eventual reality does not accept any external inputs.

14.2 Explanations of physical quantities

To highlight the strides made by eventual reality, a summary of some of the new scientific explanation and interpretations is given.

14.2.1 Existence of scientific laws

Scientific laws describe the coherent quantitative symmetries.

14.2.2 Dimensionality of space-time

Space is eight dimensional with eight branches leading to an observational space of just four dimensions.

14.2.3 Continuity and uniformity principles

Perceived smoothness across space and time, results from collective averaging over scales that are much large than the Planck scale. This smoothness produces the smooth continuity of variables and the temporal uniformity of laws.

14.2.4 Determinism and causality

Determinism and causality result from the certainty and classicality of huge random ensembles.

14.2.5 Phase space of statistical mechanics

Phase space is the transformed part of the true mathematical-complex structure of 8-dimensional hybrid space.

14.2.6 Second law of thermodynamics

The second law of thermodynamics is simply a consequence of the generalized form of Fermat's principle.

14.2.7 Quantization and discreteness

We generate a richly informative description from a spatially resolved description. A spatially resolved description requires constructing local representations.

14.2.8 Compatibility of quantum measurement rule and time-evolution principle

Quantum measurement and time evolution of the quantum state effectively exist on different storeys of reality. Not unrepentantly different processes come into play in describing them.

14.2.9 Quantum measurement

Quantum measurement selects an undetermined result by reflection of unwanted components backwards in time.

14.2.10 Symmetries of fundamental particles

Particle symmetries, as revealed by quantum numbers and statistics, are fixed by the aethereal symmetries.

14.2.11 Algebrae of quantum matter

The transformation to a collective reality into a system of local elements introduces Grassmann algebra.

14.2.12 Actuality of wavefunction

The wavefunction on a branch is a true quaternion mathematical-complex probability.

14.3 Deductions of eventual reality

Eventual reality has made deduction about the structure of space, the symmetry of fundamental particles and the algebra of quantum mechanics.

Epilogue to part 2

Current theories of reality are deficient in ways that eventual reality clearly is not. The theory of eventual reality is not complete yet but the theory is highly successful in making some physical predictions. The theory intrinsically includes the mind as necessary. Ockham's razor judges the hypothesis to be highly elegant. It is philosophically self-consistent.

PART 3:
SUPPLEMENTING THE PARADIGM WITH AFFINE CURVATURE EXPRESSIONS

Prologue to part 3

Eventual reality cannot be considered to be fundamental unless it can also be shown to generate general relativity. We will use the methodology of the existing map to accomplish this aim.

CHAPTER 15 : STATUSES OF THE THEORIES

We will discuss the status of and accompanying difficulties with both the approximate form of eventual reality that was proposed in part 2 and the standard general relativity.

15.1 Critique of eventual reality

In part 2 we ignored the curvature of space thereby excluding gravity and realist cosmological models, from the theory. To date most fundamental properties have not been computed. The presented circular logic is unconventional and may be contestable. The universal langue aspects of the theory remain very primitive.

15.2 Status of classical general relativity

General relativity is steeped in classical ideology. The coordinates of space are introduced as rank-one tensors over mathematical-real numbers. The algebraic representation of tensors is justified empirically.

15.3 Flat-space metrification

Metrification requires consistent transportation of scale and direction.

15.3.1 Transport of measures for tensors

A bilinear product of rank-one tensors is justified empirically. The product is made to fit the requirements for the 'signature' of space.

15.3.2 Flat-space interpreted by eventual reality

Provided that the metric factorizes appropriately, eventual reality can transport a scale length. However, the flat-space signature of coordinate space is Euclidean not Lorentzian. We note that the tensor algebra cannot describe quantum matter.

15.4 The metrics of flat-space eventual reality

The collective transformation that delivers quantum mechanics gives hybrid space a Lorentzian signature.

15.5 Understanding reduced spaces

We cannot parameterize a curvature in the mathematical-complex factored domain. Hence we explain the relevance of a reduced 4-dimensional space. We confirm the transformations of special relativity with a discussion of the twin paradox.

15.6 Connexions

Vector transport is quantified by defining a covariant derivative. Expressions are given for tensors of different rank and weight in terms of the affine connexion.

15.7 Tensor characteristics

Two useful tensor parameters are defined: the tensor weight and the coordinate power.

CHAPTER 16 : EXPLORING MAPS

We expect that the map can be used to help take general relativity to a lower reality. However, first we reconfirm the general utility of the map by reinterpreting the transition from classical mechanics to general relativity.

16.1 Classical mechanics to general relativity

The map should not need modification if it can also be shown to be useful in the other great twentieth century shift in the scientific worldview.

16.1.1 Step 1: map

We adopt the map and its beliefs essentially unaltered for the transition from classical mechanics to general relativity.

16.1.2 Step 2: dark clouds

Classical mechanics does not explain the equivalence of inertial and gravitational masses, the Newtonian paradigm or the precession of the perihelion of Mercury. This first difficulty is identified as the dark cloud issue. We add a principle of the universal equivalence of gravity and acceleration.

16.1.3 Step 3: master principle

We generalize the added dark-cloud principle to an equivalence of all frames: the view from every frame of reference is to be of equal worth as far as laws of nature are concerned.

16.1.4 Step 4: duality

Both the previous two steps point towards the relevance of the duality of space and matter to the new theory.

16.1.5 Step 5: environment

Historically field theories were making inroads into many areas of physics including the fields of electrodynamics, elastic materials and fluid flow. Geometrical theories of space lend themselves to field-theoretic analyses.

16.1.6 Step 6: assembly of picture

The identified conclusions from the steps of the map can be assembled into a recognizable sketch of the theory of general relativity.

16.1.7 Map summary

The steps of the map are summarized. The steps give a simple breakdown of initiators from which general relativity is clearly derivable.

16.2 First steps beyond general relativity

General relativity retains many classical precepts suggesting that general relativity is further away from the foundations of science than quantum mechanics. To generate a different understanding of eventual reality, we propose to start our next map-based journey from the macro worldview centred upon general relativity.

16.2.1 Step 1: the map

The evidence suggests that we may deploy the same map that we used on previous sketched journeys.

16.2.2 Step 2: dark clouds

General relativity does not fully describe matter nor does it describe all of matter's interactions. General relativity is causal and deterministic. The philosophy of general relativity is incompatible with the philosophy of quantum mechanics. We overcome these problems by adding the mathematical structure of space and matter as described by eventual reality.

16.2.3 Step 3: master principle

In step 2 we added the rationally justified algebra from part 2. We generalize this addition by insisting everything has a justification.

16.2.4 Step 4: dualities

The earlier steps suggest that we look to the duality of deterministic exactitude and statistical imprecision. The reformulation should start from the less-precise concept.

16.2.5 Step 5: environment

The part-two view of eventual reality is a strong ally.

16.2.6 Step 6: picture

Hence we sketch a picture of a statistical reality lying beneath the recognizable space-based form of general relativity.

CHAPTER 17 : RELATIVISTIC FORMALISM OF EVENTUAL REALITY

We aim to present an unadventurous curved-space version of eventual reality in its simplest form by examining the classical limit.

17.1 Lorentzian space

Firstly we consider flat space. An extremal variation of particle fields gives the analytical form for the entropy-parameter density, Λ. An extremal scaling variation of the metric gives the equation of motion, $\Lambda = 0$.

17.2 Riemann space

We examine a reduced four-space and add a term for the information from spatial curvature. The extremal condition delivers Einstein's tensor equation of general relativity.

17.2.1 Geodesics

Examining particle geodesics allows us to confirm that this equation approximates to Newtonian gravity and confirms the size of G.

17.3 Non-Riemann space

In eventual reality, practical units (rooted in matter) are time dependent. Necessarily we introduce an expanding scale and hence a cosmological-repulsion term is appended to the fundamental tensor equation of general relativity.

17.4 Cosmology

Our picture of cosmology is very different from the conventional material-monism picture. We will only consider an introductory cosmology here.

17.4.1 Space fatness and matter scales

The transformation relations require that space be macroscopically and microscopically flat.

17.4.2 Matter homogeneity and discreteness of space

The large-scale uniformity of matter and energy through the universe is unrelated to any prior thermodynamic contact between regions. Such properties are simply age-related quantities. The different spaces use different units to represent information and hence the spaces scale differently.

17.5 Hypothesis revealed

One equation rules our quantitative world. Fermat's principle is the first principle of the quantitative universe.

Epilogue to part 3

Eventual reality is remarkable in that it is the only parameter free, self-authenticating, multiple storey explanation of the foundations of science, philosophy, logic, mathematics and mental phenomena.

Finale

The theory has been presented as clearly as I am able. Inevitably it will meet resistance from many sources. It has much to say about many things beyond science. Thus I hesitatingly conclude with nine questions and answers about the compatibility of religious belief and eventual reality.

Preface

New ideas call for new words or for old words to be used in new ways. Thus, to help make this work easier to access and understand, the author has assembled a list of some important neologisms that are used in this work. The new terms are mostly introduced in part 2. Many words can have multiple interpretations. Hence explanations of some other words have been included to emphasize how they are to be interpreted in the context of this work.

Aether or Aethereal field	A structure holding residual information about the stubble of coherent coordinate space. The aethereal field is qualitative, describable by the universal langue and it exists in different forms in both coordinate space and hybrid space.
Apeiron	The boundless infinite system of monads that comprises the timeless fundamental universe. The apeiron has no sense of place or time; yet, it has infinite potentiality.
Branches	Coordinate space and derivative spaces have eight branches that espouse different coordinate symmetries. Separate branches are joined across branch cuts.
Coordinate space	The most fundamental structure that represents a topological space. Coordinate space exist on the first-elevated storey of the edifice of reality. Coordinate space has eight dimensions and eight branches.
Event	Events are monads with the single capability of being pointers to other events.
Eventual Reality	A comprehensive hypothesis of reality that is founded upon an apeironic network of events, that is derived from self-authenticating tenets, that presents an a priori explanation of all storeys of the edifice of reality.
Existence	Membership of the category of entities that are a conceptual part of reality.
Hybrid space	Hybrid space is constructed by transformation of coordinate space. Quantum mechanics applies in this space; particulate matter exists in this space. There are different forms of hybrid space corresponding to different unit systems.
Langue or universal langue	A language system that has a one-to-one correspondence with the permissible range of qualitative structures representing the aethereal field.
Monad	An instance of the unique primal form on the foundation storey of reality. Only monads fundamentally exist.
Natural Philosophy	The discipline that primarily studies the conceptual foundations of the objective natural world covering both

	the quantitative (as described by the sciences) and the qualitative (including the mind). It also covers the foundations of other ancillary academic disciplines that provide tools for indirectly examining nature, such as mathematics and logic.
Network of events	A network formed by the pointer links of events. The network comprises the foundation picture of the universe.
Noumenon	A process occurring within the mental or aethereal realms and amenable to mental reflection. Ideas, concepts, theories and worldviews are noumenal. This aethereal realm is more expansive than the material realm and exists along side the material realm.
Otherwhere	Coordinate space introduces a concept of space for some but not all events. Thus events can be somewhere or nowhere. Somewhere and nowhere are conveniently grouped and described as otherwhere.
Phenomenon	A real occurrence or process in an exterior world beyond someone's mental perceptions and amenable to direct or indirect scientific experimentation or observation.
Pointer	A directed referential link between events. The link possesses a sense (ie from event A to event B) but the link has no inherent length or inherent topological direction.
Reality	The totality of aspects of the universe that are posited by a set of theories associated with a common worldview of either science or natural philosophy.
Science	The set of substantive academic disciplines that describe and predict all aspects of the quantitative universe of observable phenomena.
Storey	A storey is one tier of a hierarchy of descriptions, making up reality. Each storey of reality is described using a different set of concepts and precepts.
Stubble	The stray pointer functions of events, in coordinate space, that do not participate in the dominating sub-network describing coordinate space.
Sub-network of events	A conveniently chosen selected part of the network of events.
Truth	That which conforms to the framework of belief held by science and is not contradicted by any direct observation.
Universe	A title for the unique boundless set that holds all with the sole exception of itself. It encompasses things and processes; the material and the immaterial; qualitative and the quantitative; facts, ideas, concepts and understandings; multiple storeys of reality and science and nescience.

In different parts of this work, several other common words have been used to convey their distinctly different technical and non-specialist meanings. These common words include the linguistic terms: complex, real, imaginary and proper. This work attempts to provide a proper meaning to the complex reality of our experiences. Thus to minimize the scope for misunderstanding, when such words are used in a technical sense they are given a hyphenated qualifier eg mathematical-complex and relativistic-proper. A few words can have multiple technical meanings, eg dimension can refer to a unit-dimension or a spatial-dimension. Context will not always differentiate these uses and hyphens are more sparingly used. It has been the author's intention to make this work as accessible as possible; thus I apologize to any readers who find this unconventional verbosity annoying.

Part 1:
Evolution as Vehicle for Elucidating Novel Theoretical Syntheses

Prologue to Part 1

Every generation ponders the ultimate question of 'life, the universe and everything', and collective wisdom is enriched. However, the ultimate answer has eluded us for unclear reasons: maybe we have tried too hard; maybe the question is too imprecise; maybe the existing paradigms in each era have hindering components that have proved too powerful to ignore.

Inductive reasoning would suggest that we try another tack to unravelling the ultimate question. Thus (in part 1) we shall propose that we should consider a solution for the penultimate question: *How do we solve the ultimate question?* This reformulation of the goal is not mere wordplay. Just as a word differs from the meaning of the word, science differs from the universe that science explains. The change of focus afforded by the question recognizes that science (and only indirectly the universe) determines what we call reality. Rather than considering new explanations of the universe; in part 1 of this document, we shall consider how science manufactures breakthroughs and paradigm shifts. Then, we shall construct a new worldview by emulating these processes and by anticipating the result of the next dramatic shift in our understanding of reality.

The processes of science have been studied by sociologists, by philosophers, by historians and occasionally even by scientists. Clearly science cannot provide a definitive deductive answer to the penultimate question – this question is not amenable to scientific study. However, we can uncover the persistent often unexpressed beliefs that regularly shape the formation of new scientific principles and even scientific worldviews. It is useful to imagine that these beliefs are part of the evolutionary pressure that helps to select new scientific worldviews.

Thus in chapter 2, a mix of these beliefs and historical scientific experience is used to sketch a stepwise-procedural handbook guide to accomplishing major evolutionary changes in scientific paradigms. This guide (called the map) provides one solution of possibly many solutions to the penultimate question. Then, these steps (up to step 6) are explored in the subsequent chapters of part 1, until the beginning of a new worldview comes into sight. The final two steps of the map (delivering a workable hypothesis) fall into the domain of part 2 of this work. To complete this picture, these steps are repeated in part 3 from a different starting point.

The proposed methodology for solving the penultimate question, does not guarantee that the outcome of the analysis will be useful. Nevertheless the result of these conjectures is a picture of a reality that should be worth considering and worth scientifically testing. Thus this picture of reality as first sighted in part 1, will be turned into a workable theory in the latter parts of this work and the validity of the theory is investigated there.

We will discover that the picture, which we glimpse in part 1, is a radical departure from the conventional scientific worldview. At first this newly developed picture may seem strange; the picture reveals a fundamental universe that is entirely immaterial. This picture holds that ultimate reality is an infinite sea of possibilities expressed by a qualitative network of elementary pointers. The pointers are termed events; the network is called the apeiron. This apeiron is not quantitative; it is without shape or size. Yet despite these seemingly intangible features we are able to link this immaterial picture back to a precise recognizable quantitative reality.

However, let us start our investigation with a consideration of the penultimate question.

Part 1

Chapter 1 : The Search for Meaning

1.1 Introduction

Part 1 of this work is laden with many unprovable ideas. Thus part 1 is intentionally thought provoking; it is intentionally insightful with more than enough incitement and possibly excitement. The current-chapter title hints at the central motivator for the work reported here. We are driven to ask and try to answer questions such as: What is the nature of reality? How is it that we can understand the universe? What are the determiners of science?

The long-term role of science is to provide this understanding of nature. Although it is rarely denunciated, current science is in poor shape. We have become so accustomed to apparently only having a few known hurdles to overcome that we presume that our minor failings in understanding are correctable by slight adjustments to our current models rather than adjustments to our founding principles. Note that on the large scale, we don't understand the size of the universe, dark matter, dark energy or cosmic inflation. On the small scale, we don't know what the smallest constituents of matter are (eg particles, strings, branes) or how many dimensions there are to space. On the intervening scale, science cannot accommodate the only thing that any of us truly knows to exist, namely our mind. Then there are the dilemmas of quantum measurements and the issues of combining gravity and quantum mechanics. These inconveniences could be small independent problems for our models (that describe what constitutes nature) or these inconveniences could be one major philosophical disaster for our paradigm of reality. It is always hard to relinquish a familiar worldview that has explained so much. It is equally difficult to build up a replacement worldview when our thinking is dominated by a highly successful previous worldview.

Now, inquiries into the ultimate questions of scientific endeavour are hardly new. Indeed so much has been written and debated by so many, for so long, across so many cultural specialities including fiction and comedy, that any review or even credit for originating individual ideas is likely to be quite inadequate. For instance, we know of many speculations by ancient philosophers of many cultures. However, the philosophers soon discovered that there seems to be a counter argument for every assertion. Such old work that has previously led nowhere does not provide convincing guidance to the best methodology for answering the truly fundamental problems.

We could attempt to build an argument based upon what we feel the universe ought to be like (eg mathematically elegant). However, it is impossible for us to ascertain how biased such a view would be. Hence, the only approach with a proven track record is the one that follows the scientific method of asking what features exist in the universe and how these features are manifested. This method relies upon inference, which is sourced from repeatable observation, to instruct us.

In part 1, we do not attempt to answer the ultimate query directly – later parts will deal with related questions and will provide answers. In part 1, we start by asking: How does science develop new pictures that could answer such fundamental questions? In the present climate of opinion imposed by the current paradigm, this indirect question is easier to discuss than questions of ultimate truth. This easier question is one that redirects the focus away from a study of nature towards a study of science.

So what is the next major achievement in the evolution of science? Of course this question is too imprecise since the question has many possible answers. However, the question will dominate part 1 of this work. Many readers will find the solution that we shall present challenging.

For the most part, science proceeds by slow accretion of knowledge, of understanding and of belief. Thus science has a structure that grows and evolves and, in the process, science both adapts to and helps to reform cultural beliefs. However, very occasionally, sometimes tragically (eg Giordano Bruno[1] and Galileo Galilei[2]) and often with great reluctance, conventional wisdom relinquishes old cherished beliefs and science makes a spectacular shift in its recognized founding principles. In the twentieth century, our forebears witnessed the dramatic impacts of relativity and quantum mechanics on our understanding of the nature of reality. Such radical jumps in our understanding of nature are necessarily rare and difficult to accomplish. Nevertheless the purpose of part 1 of this work is to describe how we might possibly herald the next spectacular stage in this scientific progression.

Scientific investigations can be like expeditions into unchartered territory. Here an investigated query is analogous to the expedition's mission objectives. Our primary query (mission objective) is to uncover the nature of reality. Yet this request is a little like asking a pre-industrial ancestor to go from the delta of the river Ganges to the summit of Mount Everest without proving any help or any further instruction. A random walk (that does not traverse much of the Eurasian land mass) would be unlikely to get our ancestor to the desired goal. Both of these aforementioned physical and philosophical tasks would be greatly assisted by the provision of a map (with instruction as to its use). Such a map serves two important purposes. Firstly the map gives direction to the journey avoiding unnecessary excursions. Secondly the map provides milestones that flag how much of the journey has been achieved and how much remains to be accomplished. Without these milestones, one can easily image that some dispirited travellers (perhaps suffering from the cold and lack of oxygen) could give up the quest when they were unknowingly very close to their goal. Thus for any prolonged and arduous task, a good map is an almost essential aid for enabling success. Obtaining a map will be our first objective.

1.2 Guide Map

The map is to be regarded as a series of instructions which when followed will suggest a new more fundamental paradigm. The first step in this task, and therefore the first step of the map, involves acquiring or sketching the map. Rationalists may snipe that maps are unavailable for previously unvisited and unobserved destinations. However, this is simply the reason why we must infer details of the map from prior experience of earlier journeys. Note that the processes and steps by which we go from an old picture of reality to a new picture of reality may have necessary similarities. Hence, in an indirect procedure, the map can be said to have been developed from observation. This development is imbued with a series of beliefs about science. These beliefs are drawn from the myths of science. Thus in this part of this work we shall construct a set of inferred steps (defining the 'map') that are mostly generated by retrospectively reconstructing the last major paradigm shift, the transition from classical mechanics to quantum mechanics. Then we shall follow this analogous set of steps on a journey from our existing paradigm to a new description of reality.

We can view science as being a creation of human culture. Within this culture, stories loosely based upon fact, are told about how science has developed over the ages. Some stories have become 'urban legends'. Indeed, who has not heard the stories of an apple[3] falling from a tree or an excited naked scientist[4] running through the street? This collection of allegorical stories is used to teach how the ideas of science can be developed. Generally without acknowledgement, these evolving cultural myths of science have provided an instruction book for the creation of our analytical understandings of everything from individual scientific experiments to the scientific principles. With one further layer of abstraction, we shall invent the tale of the creation of these stories of science. Effectively this creation tale describes a methodology for the *advancement* of the scientific paradigm for describing reality. This creation tale is the instruction map for advancing our understanding.

In developing this creation tale, certain meta-principles will be followed. By meta-principle, I mean a guiding belief that is likely to apply to most, if not all, physical principles. Here we will only be concerned with those meta-principles applying to paradigm shifts. Such meta-principles describe one aspect of the belief structures underpinning this work.

The first meta-principle that we shall apply will be referred to as the *theory of similar form.* This meta-principle is the basis of the argument that we have just discussed, for the development of the map. This meta-principle inherently recognizes that the development of science is a human intellectual activity that is governed by the principles of evolution. We expect that scientific development will obey evolutionary principles. Thus science reports an accumulation of novel ideas (some heretical, some unproductive, some counter-intuitive, some paradoxical) and science reports gradual refinements of all these ideas until non-ignorable competing requirements (from new circumstance or more effective competitors) force a change of the mainstream worldview. Thus, over time, when we are faced with similar difficulties about how to understand the universe we, like the biological version of natural selection, are likely to respond collectively with similar strategies. In particular, conceptually recent, challenging, evolutionary environments (that have previously engineered a change to the paradigm) are likely to exhibit similarities and thus, at each revolutionary break, the processes of natural selection may follow similar evolutionary paths. Hence, to progress beyond quantum mechanics, we will follow a similar progression of steps to a fabled progression that led us to the advent of quantum mechanics roughly a century ago.

The design of this progression of steps is to be moderated by the second meta-principle. The second meta-principle is termed the *theory of historical revisionism.* This principle allows us to reinterpret those earlier events to fit what seems to be a logical development. We are not interested in our or nature's mistakes (evolutionary dead-ends); a believable success story is more useful than any adherence to accurate fact (and particularly chronology). Thus we shall not let the accidental manner, in which history sometimes develops, get in the way of an expansive predictive argument based upon how it most likely could have developed in a more logical way. In other words, in this context, an illusory guided design is more instructive than any rigorous historical truth.

The third meta-principle will be referred to as the *theory of the difficult change*. The basis for this meta-principle is the belief that if by tinkering with our descriptions of reality (i.e. making refinements here and there) the desired change could have been brought about, then that change in our theories would already have been proposed, tested and accepted. Therefore we should expect and only seek out big changes in our conceptual foundations. For instance, we argue that after a century of moderate development to quantum mechanics those adaptations that are currently needed are major changes that will require a new paradigm. This is a powerful belief; when we are faced with two choices about how a solution should be developed, one choice involving a major shift in belief and one choice a minor shift, we will argue that we should accept the more radical major choice provided that the choice satisfies the fourth and last meta-principle.

The fourth and final meta-principle will be called the *theory of fresh eyes*. This meta-principle is a particular form of Ockham's razor. It is a mix of *simple is best* and *be humble before nature.* Inescapably we have a vested interest in our current belief structure; this belief structure has served us well and only seems to have a few potential problems. Nevertheless, embracing the possibility of simpler alternative beliefs may take us forward, yet clinging to old productive inflexible beliefs can only hold us back. Thus we will be prepared to reinvent challenges for current science and redescribe them in different formats. We accept how nature is observed to behave but ask whether we are interpreting these behaviours correctly. Are there other simple and elegant worldviews that we should consider? We shouldn't take anything for granted; sometimes we can be beguiled into refusing to accept an obvious explanation. The conventional belief may not be the simplest and we must not try to impose our historical prejudices upon nature.

Thus, by calling upon the theory of similar form, we shall form a map by looking at our recent experiences with changing scientific worldviews and by applying the four meta-principles. The procedure for generating the map is quite simple. In principle, this procedure can be repeated *ad infinitum* with refinement. This identification of the questions that comprise the steps on the map is regarded as the first step of the journey. An accomplishment of this step is described in this chapter. So if you are sitting comfortably, then let us begin.

Once upon a time, at the turn of the last century, physics was in apparently good shape with substantial experimental confirmation of basis principles and just a few inconsistencies between some predictions and experience. Understanding how to resolve these 'small dark clouds[5] in the clear blue sky of physics', led to the revolutions of relativity and quantum mechanics. Prior to eventual reality, science is apparently in fantastic shape and there seem to be no clear inconsistencies between experiments and theory. Our theories are better tested than ever before. Nevertheless there are incompatibilities between different major theories and even internal interpretive incompatibilities within major theories. Furthermore, science has boarders: many secrets of the universe are beyond current science. These issues suggest that the next abrupt advance in science requires, first and foremost, not simply an improved formulation and consequentially a new understanding of the material world but a radically new reinterpretation of the universe and consequentially the advance will require a significantly modified reformulation of our entire thinking about reality.

1.2.1 History Rewritten

Before we can write the next steps in the map, we need to consider how to repackage the historical development of quantum mechanics into a generic form that we shall be able to apply to other worldview transitions. Thus let us briefly review some of the stages in the gestation of quantum mechanics from inception to birth.

Quantum mechanics had a painful birth and difficult pre-natal development, with many famous scientists lending a hand. Max Planck was responsible for the conception of quantum mechanics and he is rightly referred to as the father of quantum mechanics. Planck inseminated classical mechanics with foreign material: the quantum idea. Hence classical mechanics was the mother of quantum mechanics. Earlier, Raleigh and Jeans had noted that classical physics could not explain the spectral form for blackbody radiation. After five years of study, Planck, effectively resorting to the theory of fresh eyes, came to the inescapable conclusion that energy (in the form of electromagnetic radiation) was emitted or absorbed by electron states in discrete quanta. He published this work in December 1900 at the dawn of the twentieth century and the story of quantum mechanics was begun.

Einstein, who was not fazed by challenging ideas, helped to further grow the quantization concept. Einstein showed how the photoelectric effect, which involves the ejection of electrons by solids, could be explained by using the same principles. Then, he showed how the low-temperature behaviour of the specific heat of solids could be explained by requiring that the vibrational states of solids be quantized.

The rapid strengthening of quantum concepts was nurtured as new problems were encountered for classical mechanics with Lord Ernest Rutherford's description of the internal structure of an atom. The atom was observed to be stable despite the anticipated requirement, in classical mechanics, that orbiting (ie accelerating) electrical charges must radiate electromagnetic energy. Even though in 1913 the quantum theory was still only partially formed, quantum mechanics now became quite prominent as Niels Bohr explained the stability of atomic electron states and their line spectra by applying quantum principles to Rutherford's atomic model. The hypothesis' strength grew further with the successful explanation of Compton scattering. Then, in 1924, Prince Louis-Victor de Broglie triggered the onset of labour for the birth of quantum mechanics. He preposterously proposed that the wave-particle dualism should have universal applicability.

Quantum mechanics was delivered twice in 1926, first by Werner Heisenberg and then by Erwin Schrödinger. However, it was soon shown that it was the same infant theory. The following year at the Solvay conference, quantum mechanics was baptized with Niels Bohr as her Godfather and quantum mechanics was accepted by the scientific community despite some very disturbing issues of interpretation. These problematic issues of interpretation remain intact even though today's quantum mechanics is significantly different from those early forms. Among many others involved in guiding the development of quantum mechanics, Paul Dirac was a key tutor who greatly encouraged and stimulated the infantile development of quantum mechanics. Dirac pioneered at least three significant contributions that were fully developed by others; these contributions include his work on the introducing relativistic principles, his work on the path-integral form of quantum mechanics and thirdly his contributions to second quantization, a reformulation that led to quantum-field theory. Each of these advances will have further mention later in this work.

Next, as an example of the impact of a paradigm shift, we shall discuss the Copenhagen interpretation of quantum mechanics. Then, the historical steps in the development of quantum mechanics are recategorized so that the transition might be emulated in a new switch of our fundamental belief structures.

1.2.2 Copenhagen Interpretation of Quantum Mechanics

There are rival interpretations of quantum mechanics; however, the Copenhagen interpretation[6] of quantum mechanics is the orthodox interpretation that is accepted by most physicists. Minimalism is the key strength and key weakness of this interpretation. The interpretation has delivered a comprehensive workable picture by attaching a minimum of explanation to the mathematically predictive quantum formalism.

In the Copenhagen interpretation, the quantum state is a conceptual holder for all the information that characterizes a system. Reality is directly sourced from the classical-mechanical world; thus quantum reality is attributable only to those features which have been directly measured, which have been indirectly measured or which are otherwise in the classically real realm. The quantum state (and its spatial representation known as the wavefunction) is not regarded as being a physical entity. Thus it does not matter that when a measurement is made, the state of the system abruptly, instantaneously and irreversibly changes defying relativistic expectations for anything physically real.

Now, the paucity of attributed meaning is without explanation and thus some researchers have attempted to go beyond the standard minimalism and these researchers add further hypothetical detail. In different approaches, such further details have included an underlying definite reality, an active role for the observer and various systems of co-existing parallel worlds. However, there is no supporting evidence for any of these alternate descriptions. Furthermore, each attempted new interpretation carries its own unsatisfactory burden. It should be noted that entanglement, non-locality and the impracticality of maintaining a classical-style reality between measurements seem to have been successfully confirmed experimentally. Hence it is an imposing task to attribute to quantum mechanics more than a minimum of interpretation whilst retaining plausible and consistent explanations of phenomena such as entanglement, non-locality and complementarity.

Quantum mechanics takes a small step away from classical certainty. The definite classical reality in which we can attribute an enduring existence to features and properties of matter is rejected. The attributed understandings of the quantum paradigm can be placed into four categories: 1) the nature of reality, 2) the core principles, 3) the mathematical formalism and 4) the characteristics of applications.

Let us consider these four categories in reverse order. The characteristics of applications include the principle of superposition and the principle of conservation of energy-momentum. These principles are undoubtedly important yet they depend upon the ancillary model input (eg the form of the interaction) providing the technical description of the system to which quantum mechanics is being applied. In this set of categories they are deemed less revealing of the quality of true reality. Now, the next category, the style of the mathematical formalism, describes the nature of the defining equations of quantum mechanics. Thus one cannot materially change this formalism without adopting a new theory. The next category holds the core principles. The core principles include the principles of complementarity, uncertainty, non-locality, entanglement, causality and determinism. Finally the nature of reality provides the key links between the quantum description and perceived reality.

Now, the first three categories provide the general scope for interpretations in quantum mechanics. This scope has been highlighted by a series of questions that are presented in table 1.1. A novice to quantum mechanics would probably prefer a comprehensive series of definitions, principles and features. However, the questions in the table should suffice for the reader who has had some prior exposure to quantum mechanics.

These questions are grouped according to categories. Thus questions 1 and 2 deal with the nature of reality; questions 3 to 8 cover the core principles and questions 9 to 12 describe the mathematical formalism.

Table 1.1

The table provides a comparison of the interpretations provided by classical mechanics (here incorporating special relativity) and quantum mechanics (based upon the Copenhagen interpretation) describing the dynamics of matter in space.

	Question	Classical	Copenhagen
1	What is measurement?	Determination of knowledge about system	Linked association of system to a classically real variable.
2	What is real?	Anything that can be measured.	Knowledge acquired from measurement.
3	What is consequence of a perfect measurement?	Nil.	Scrambling of complementary variable.
4	What is the spatial range of measurement?	Range of causally chained influences of measurement.	Full range of system.
5	How quickly do effects of a measurement propagate?	Speeds up to speed of light.	Instantaneously.
6	What happens during a measurement?	A predictable result is observed.	The system selects an outcome weighted according to the square

			of the modulus of the wavefunction.
7	What happens to a system between two measurements?	It changes analytically between states corresponding to the two results.	The wavefunction evolves analytically; the system is unspecified.
8	Using perfect information, what can we predict?	Everything about future and past.	The probability of future outcomes of experiments.
9	What algebra describes evolution of material systems?	Mathematical-real.	Mathematical-complex.
10	To what is matter fundamentally reduced?	Point particles.	Quanta.
11	Are fundamental units of the same type distinguishable?	Yes.	No.
12	How does matter interact with other matter?	Via forces at a distance.	Through local-field interactions.

1.3 Dark Clouds

Near the end of the nineteenth century, according to what has become folk law, the extremely successful worldview of classical mechanics was compared by Lord Kelvin with a clear blue sky in which only a few small dark clouds obscured some difficult-to-ascertain truths.

Scientifically dark-cloud research is always fascinating and it wasn't long before Kelvin's nominated dark clouds shattered the classical façade of reality. The term, dark clouds, can be used to refer to those problems that serve to obscure particular scientific areas in which our traditional theories ought to work but inexplicably fail. Dark-cloud research is the substance of scientific legends. Many famous gedanken experiments are examples of dark-cloud studies. For instance, Galileo[7] imagined dropping different-mass objects from near the top of the leaning tower of Pisa and Einstein imagined travel on a Swiss tram at relativistic speeds. At the time both of these imaginings exposed inadequacies in the then dominant descriptions of reality. However, new experimental evidence can equally generate a dark-cloud issue. Studying such inconsistencies may be called dark-cloud research. It was another dark cloud, the aptly named ultraviolet catastrophe, which led Planck to propose that radiation was emitted and absorbed by atoms in discrete quanta. Planck's suggested explanation marked the start of the transition from centuries of classical thinking to the Copenhagen interpretation of quantum mechanics. In a similar way, we shall look to step two of our map, to mark the beginning of a new journey to a more fundamental level of reality. Hence by analogy with the initiation of quantum mechanics, we will require that step two on our developmental journey should involve identifying what it could take to resolve the conflicts inherent in the deficiencies (lurking dark clouds) of quantum mechanics. In particular, we are to be more concerned about what new unexpected concepts and precepts we may need to add to our theoretical description rather than what features we may need to abandon.

Many conceptual problems abound within our twentieth century descriptions of reality. Some of these problems are known and researched, others tend to be overlooked or even denied. It is often not possible to tell whether our principles are wrong or whether there are missing extra model features that are still to be confirmed experimentally.

Each step in the map is to be quite manageable. We should not overburden the instructions. The following remark may help to clarify the role of this stage of the map. The scientific community can be self-congratulatory when it looks at how much of the universe is explainable by science. Almost by definition, we call that (blue-sky) part of the universe that is explainable by science, the physical universe; alternative names for this realm of science are the material world or the quantitative universe. However, the experiential universe, as we intimately know it, is much more relevant to us personally. Largely, much that is qualitative is unexplained and indeed the mental realm is quite unexplained. We hope that our new theory will explain such matters. However, such issues do *not* belong to our set of dark clouds. Dark clouds relate only to problems with existing theories as applied to areas in which the existing theories are clearly deemed to work ie the physical universe.

Thus the second step on the journey involves answering the question: What new foreign concept and/or precept could we append to the existing description of reality that would remedy the perceived difficulties that arise from attempting to apply the existing descriptions across their supposed area of applicability.

We shall complete the instruction of the second step in chapter 2. However, in the next section, we shall consider the instruction for step 3.

1.4 The Master Principle

Now, it took about a quarter of a century, from the first suggestion of quantization to the first restricted (non-relativistic) formulations of quantum mechanics by Heisenberg and Schrödinger. It will be noted that Planck's original quantum hypothesis was very limited in its suggested scope.

However, following the work of Einstein and others, it was gradually realized that the notion of quantization might be applicable much more widely. This wider application was not just that of extending the application to Bohr's atomic model or to extending the application to more complicated molecules. It was the appreciation that the principle of quantization was an overriding 'master principle': a principle that should be applied to every property of every thing in every circumstance. Identifying and accepting the master principle was an important step in the development of quantum mechanics. The master principle has given quantum mechanics its name.

During the intervening quarter century of the development of quantum mechanics, there were also many other clues to other aspects of theory. However, these clues, which relate to the derived characteristics of the new theory, will be dealt with in later steps of the map.

Thus, by analogy with the development of quantum mechanics, we assign to step three of the map an instruction to identify the overriding master principle for the new theory of reality. Note that this master principle does not tell us about how to formulate the new description of reality but it does tell us something crucial about the core belief structure of the new formulation.

Step 3 is to follow from step 2 of the map. Recall that step two of the journey identified some essential new fundamental elements to be appended to our description of reality. Specifically the third step will ask that we generalize one of these new elements (taken from step 2) in the description of reality. We are to expand this element into a master principle. Thus the role of this third step is to identify and define the master principle by investigating generalizations, of those new elements of step 2, that we could sensibly apply to reality.

1.5 Dualities

Within the concepts of classical mechanics the ability of elemental matter to appear as either particle-like (when detected by a photographic plate) or wave-like (when diffracting) is a complete contradiction. Not surprisingly this classical dilemma was an important guide to the construction of quantum mechanics. In general, when there are two competing concepts, such as waves and particles, one of these concepts will be dominant in the current paradigm. Thus, even though waves were important in classical mechanics, the particle concept was considered supreme; waves were generally thought to be a derived concept. For example, the ocean appears to sustain true waves; however, the ocean is not continuous it is made from particles known as molecules. Thus ocean waves are a derivative of molecular motions.

The contradictions (between competing particle concepts and wave concepts) arose when observing quantum phenomena. The resolution of these contradictions needed the incompatible concepts to be somehow unified in a different way. This puzzle of the dual particle and wave behaviours of matter is known as the particle-wave duality. Now, the theory of difficult change suggests that if by starting with the conceptual dominance of a particle picture, these competing concepts could have been suitably merged, then that merging would already have happened. Thus it is not surprising that the required merging of the concepts was accomplished (at least by Schrödinger) by starting with the previously subservient wave concept and applying the quantum (ie master) principle. Actually Heisenberg based an earlier derivation of quantum mechanics upon the concept of the quantum state. However, it was soon made apparent that Schrödinger's wavefunctions are a spatial representation of the quantum state. For the sake of a consistent tale, historical revisionism will allow us to ignore this accident of history.

We see by analogy with the development of quantum mechanics that dualities may suggest how we should begin to formulate a new theory. Thus we assign the fourth step of the map to an investigation of dualities and consequently the identification of certain characteristics in which the new description of reality should be reformulated.

Specifically this step involves:

1. recognizing the serious duality-style clashes occurring in our current description of reality
2. identifying the concept that has been traditionally regarded as subordinate in each duality
3. contemplating how a description could be based upon the subordinate concept.

Thus step four of the map involves identifying and describing the clues as to how to reformulate a description of reality. The actual picture-forming process comes in step 6 of the map. However, before we attempt this composition, we will need to take stock of where we have arrived.

1.6 The Evolutionary Environment

The previous steps of the map have identified what we should regard as important in a new formalization of reality. The map has enabled us to race ahead through fields of speculation to arrive somewhere near where we can first catch sight of our goal. This success has meant that we have avoided a lot of exploration on the way, explorations that could have distracted us. However, the map was only roughly sketched. By step 5, we should be so close to (but not in sight of) this vantage point that it is necessary to take stock of the natural scientific environment so that we can tease out the easiest route to our journey's summit.

By prescription, our new theory will involve a paradigm shift. We need to ask what competition the new theory will meet from other emergent worldviews and whether there are symbiotic alliances with other scientific theories that can be made. Furthermore, in evolution, there are often consistent developing trends within different era. Latching on to an appropriate identifiable trend can aid the development of the new hypothesis and it can provide an accommodating way of presenting the theory.

Finally just by studying the scientific environment we can better appreciate what we have come to accept about it and how to take the final steps to the summit. There are direct competitor theories and we need to ensure that any new theory is born with sufficient strength to survive its infancy.

Thus step 5 of the map asks for an analysis of which creatures in the jungle of ideas are potential rivals, which creatures are potential allies, how is the ecosystem changing and what copying strategies are currently proving successful.

1.7 Finding the Starters Block

Everything seems easier when we understand what we are doing. The map is designed to give the necessary framework for our understanding of our scientific progress and thereby make our journey easier to accomplish. However, even with the many clues already generated, the greatest difficulty in formulating the new theory comes from knowing where to begin. The results of our journey through the steps 1 to 5 will have generated the pieces of a puzzle that are to give a view of a new reality. However, we won't try immediately to formulate our new theory. We shall follow the lead of our forebears – such as Newton, Heisenberg, Schrödinger and Einstein – in recognizing how to process these pieces of the puzzle so that they will fit together and give an overview of reality. Only when we have established how to proceed, can we navigate the traps inherent in uncovering a radical departure from familiar reality.

To start the task of formulating (ie giving birth to) the new vision of reality, we ask the same question that Heisenberg did 'What do we accept as fundamentally real or existing?' Heisenberg's answer to this question was that the results of measurement are real; our answer will be quite different but equally relevant. We should crystallize our view of reality about the answer to this question.

Consequently the instruction for step 6 on the map tells us to start by asking: What is fundamental? What truly exists? Then we are to consider how the earlier steps of the map can help to create a picture of a new reality. The central entities upon which the new formulation is to be built are to be the inferior concepts of the dualities discussed in step 4. The representation of reality is to conform to the master principle of step 3. We need to also incorporate the view from step 5 so that our new theory can thrive. Hence this step aims only to assimilate the concepts and precepts uncovered during the map's previous steps, into a distant picture of the new reality. Creating the detailed formalism will be the next step.

1.8 Aiming Low

Quantum mechanics posed a challenge to its developers. Interestingly the first successful formulations of quantum mechanics were remarkably unadventurous. These formulations described the non-relativistic quantum state of a single particle in a classical interaction field. Indeed, it would have been a very much bigger leap to have either added relativity or to have satisfactorily described particle-particle interactions. Generally there are two undesirable hurdles associated with making a larger-than-necessary leap in outlook. The first hurdle to be jumped is the natural resistance by the scientific community to the acceptance of a new theory that is too radical, too much of a departure from that with which we have become familiar. This resistance was illustrated when few scientists initially took much notice of Planck's laboriously developed seminal ideas on quantization. The other equally pertinent hurdle is the intellectual effort of the developer needed to make a large leap as one giant untested step.

The further developments of quantum mechanics occurred very much later. It is commonly the case that a step-by-step approach makes an analysis easier to perform. This is why we have developed a map with a defined set of simple steps to guide us to our goal. The lesson from the birth of quantum mechanics is that the steps between the introductory and developed forms of the new reality are probably sufficiently large that the first formulation of the new reality should not be too adventurous. This new formulation may retain some of the tenets of the previous theory and may limit its scope to one not much bigger than that claimed by the previous theory. Initially the practical use of the new approach is likely to be only to those areas where quantum mechanics is found to be failing.

Furthermore, although quantum mechanics claims to encompass classical mechanics for macroscopically complex system, the transition to classical style behaviour is too complex for practical use of the quantum mechanical approach in the classical arena. Similarly we should expect that we would want to use the older theories quite extensively for some time.

Step 7 says that we should take the pictures from step 6 and reformulate reality but we should not be very adventurous. We should allow a first formulation of the theory to retain some of the remnants of old reality.

1.9 Evolutionary Survival

Science is a collaborative human activity. Science creates, adapts, forgets and rejects memes. Ideas that are not accepted within a community, whatever their intrinsic usefulness, are without functional value. Hence, it is incumbent upon an originator of a useful idea to communicate the idea to others and to present the idea in a way that aids its propagation through the scientific community.

The successful introduction of a new theory to the scientific world may take many forms. The theory will become the work of many but starts as the work of one whose work builds upon the work of innumerable predecessors. New ideas will be judged by their perceived value. The value of a new paradigm can be demonstrated in its ability to provide meaning to otherwise unexplainable principles and phenomena, in its ability to jump hurdles at which existing theories stumble, in its ability to delineate the approximate validity of older theories and in its ability to make predictions and describe new phenomena.

Thus step 8 of the map is where the details of the destination of the journey are to be documented and shared with others. Figuratively in step 8, a flag is planted upon the summit and a formal claim of jurisdiction is made in the name of the scientific community. Practically the instruction is less dramatic. The instruction of step eight is to promulgate the strengths and claims of the theory within the scientific community.

Notes

[1] Following biblical experience the persecution of (scientific) belief suppressed but ultimately strengthened the resolve of free thinkers and ensured deliberate consideration of the ideas and therefore their eventual dominance.

[2] An account of Galileo's alleged sins and of his persecution – a clash of scientific integrity and brazen political power - is given in Bronowski pp205-218.

[3] The story says that Newton, whilst resting under the shade of an apple tree, observed an apple fall to the ground and wondered why it fell downwards.

[4] The story claims that Archimedes got into a full bath and noticed that he displaced his own volume of water. He immediately recognized that he could use this technique to measure the density of irregularly shaped artefacts and was so excited that he ran naked through the streets shouting *eureka.*

[5] In 1900 Lord Kelvin gave a lecture in which he spoke of the great success of the classical Newtonian viewpoint with just two or three dark clouds obscuring the panoramic view of the clear blue sky. Ominously these dark clouds hung over the interpretation of the Michelson-Morley experiment and the shape of the spectrum of blackbody radiation (the so-called ultraviolet catastrophe).

[6] See Heisenberg (1958) for a fuller understanding of the Copenhagen interpretation.

[7] Galileo's ponderings are discussed in more detail in section 16.2.5.

Chapter 2 : The Trouble with Quantum Mechanics

2.1 Introduction

Quantum mechanics could be said to be the most successful physical theory ever devised. Since its inception around the start of the twentieth century, quantum-mechanical predictions have been supported by countless experiments. Nevertheless it is illogical to argue that quantum mechanics is the ultimate theoretical mechanics let alone the ultimate theoretical description of reality. It is not simply that every physical theory, no matter how good the theory's predictions, or how often the theory has been tested, appears to be eventually replaced by a successor; it is that quantum mechanics has some serious limitations and inconsistencies.

Despite tantalizing suggestions of a holistic nature, quantum mechanics is limited in its application, to subsystems of the universe. Quantum mechanics denies itself the capacity to be applied to a system that is unmeasurable such as the entire universe. Furthermore, the philosophy of quantum mechanics does not identify the elements of reality within the theory. Thus in the Copenhagen interpretation, quantum mechanics is only capable of relating to a recognizable reality at the instance of a measurement. The rules of measurement appear as an ad hoc unexplainable addition to the rest of quantum mechanics. These rules fail to provide an embedded explanation of the identification of a quantum state with a probability amplitude, the notion of an instantaneous change of state (often called an instantaneous collapse of the wavefunction) or the origin of indeterminacy. It almost seems as if half of the picture provided by quantum mechanics is missing. Thus, not surprisingly, there remain multiple problems when quantum mechanics strays into realms that are adequately governed by classical mechanics and general relativity.

Now, the purpose of this chapter is to transit the second step of the map. To complete this step we are required to consider which problems fit the 'dark cloud' category. Recall that the dark clouds of quantum mechanics should be those nagging problems that exist within the realm in which quantum mechanics should unquestionable hold. Then, we need to consider what we would need to add to our existing description of reality to fix up the dark-cloud difficulties (within the realm that quantum mechanics should hold).

The transition from classical mechanics to quantum mechanics was undoubtedly hindered and delayed because the transition involved the surrender of deeply held beliefs. If the transition from quantum mechanics to its successor theory, were not to involve such a denial of the 'obvious', one might also expect that the transition would already have occurred. We shall examine what aspects of quantum mechanics might need changing. We accomplish this task by partially removing a blinker and experimenting with another way, not reliant upon quantum mechanical principles, that we can interpret quantum reality. Thus for a rather special case, an imprecise formulation of a modified mechanics will be presented, one that avoids several of the problems of quantum mechanics. This formulation is analogous to Planck's original hypothesis – in that it could be regarded as identifying what needs changing to resolve the existing problems within the current interpretation of quantum mechanics rather than making a genuine break with this previous description.

For the purpose of these discussions, the new limited explanation will be referred to as free mechanics. Here the term 'free' signifies a release from many constraints. Conceptually this free mechanics is deemed to be based upon the execution of entirely incoherent proceedings. Both causality and determinism are natural casualties of this theory. In the first instance, the treatment of causality may be the hardest part of this fix for the reader to accept.

Thus through free mechanics, the observed order embedded in physical laws will be attributed to the cumulative disorder of the underlying features in a way analogous to the way that statistical mechanics predicts the ordered laws of thermodynamics from an underlying turmoil. Consequently it is demonstrated how something resembling a free-space wavefunction can be derived from a simple statistical analysis. Furthermore, the philosophical differences between the quantum-mechanical rules of measurements and the principles of quantum mechanics are removed.

The free mechanical hypothesis to be presented here is unorthodox and requires a reinterpretation of the spatial coordinate representation of reality. The first part of this reinterpretation relates to the structure of space-time and its association with measurement. The second part of this reinterpretation relates to the appropriate coordinate description of reality. However, we shall start by considering the problematic issues created by quantum mechanics.

2.2 The Measurement Dilemma

Many of the apparent difficulties of quantum mechanics are well known and these difficulties will be mentioned in this chapter. Over the years, these difficulties have lead to the development of alternative schools of thought, such as that school founded upon the concept of underlying hidden variables. These schools of thought have promoted their sets of belief to overcome the difficulties posed by quantum mechanics. The arguments presented here describe another conceptual model, one that will be used as the starting point for the picture of a new description for reality. However, in this chapter, the conceptual model has a very limited applicability and there can be no evidence that what is presented here is necessarily the correct solution to any of the difficulties. Thus what is presented should be simply viewed as a considered but personal viewpoint on the thorny question of what is the best way to overcome the conflicts of quantum mechanics. Naturally the new hypothesis requires that we give up some of our deeply held beliefs about the universe.

Now, at its core, quantum mechanics contains limitations. For instance, quantum mechanics does not define what constitutes a measurement and it can be applied only to otherwise isolate experimental subsystems that are connected to a measurement apparatus.

The quantitative formulation of quantum mechanics has two distinct parts. One part of quantum mechanics describes the temporal propagation of the quantum state; this part is an obvious descendant of the equations of motion in classical mechanics and this part shares many of the features of classical mechanics. This part of quantum theory will be referred to as the principles of quantum mechanics. The second quantitative part of quantum mechanics describes how to calculate the results of a measurement. This part contains features that make quantum mechanics philosophically different from classical mechanics. This other part will be referred to as the rules of measurement. Note that this terminology, of principles and rules, reflects the imposed-style of each of these parts of quantum mechanics. The terminology should not be regarded as necessarily giving a higher status to one part of quantum mechanics over the other part of quantum mechanics. Indeed, it is these rules that make quantum mechanics so different from classical mechanics.

A summary of the characteristics and internal conflicts within quantum mechanics is displayed in table 2.1.

Table 2.1
The table presents a simple contrast between the characteristics of the constituent quantitative parts of quantum mechanics.

Principles of quantum mechanics	**Rules of quantum measurement**
Complex & analytical	Real & non-analytical
Deterministic	Non-deterministic
Relativistically causal	Relativistically acausal
CPT invariant	CPT variant

Clearly these two quantitative parts of quantum mechanics are logically inconsistent. For instance, consider the result of a measurement on an isolated system. By virtue of the rules of measurement, the result is non-deterministic. However, now let the system under study grow to include the measurement apparatus (also assumed to be isolated from the rest of the universe). The principles of quantum mechanics would predict that the change of the state be deterministic. This logical discrepancy prevents an enforceable assignment of what constitutes measurement and even certain identification as to when the measurement occurs.

To cope with this dilemma, the Copenhagen interpretation asks for some forgiveness, as quantum mechanics is *not* a holistic theory. Quantum mechanics is only to be applied to the study of effectively isolated subsystems of the universe. Measurement equipment could be presumed to be or required to be connected to the rest of the universe. These assertions ensure that we may expand the boundary of what is the measurement apparatus, to encompass more of the subsystem under study; however, we may not expand the boundary of the subsystem under study into those parts of the measurement subsystem, that have appreciable connections with the remainder of the universe.

Nevertheless consider a hypothetical isolated robotic spaceship in distant space. If this spaceship is so connected to the rest of the universe that the principles of quantum mechanics cannot be applied to the entire spaceship, then one may ask how we can possible apply quantum mechanics to any experimental subsystem on Earth. However, if the principles can be applied to the spaceship, then we have to assume that the non-determinism of measurement is an illusion and arises from our lack of knowledge of the state of the measurement system. Then, the characteristics of the principles of quantum mechanics always fundamentally hold and the unacceptable philosophical inconsistencies of classical mechanics still hold for quantum mechanics.

Such arguments are part of a third constituent part to quantum mechanics. This third non-quantitative part is its philosophical interpretation. Note that only the quantitative parts are available for falsification by experimentation. Hence, it is impossible to unambiguously choose between alternate philosophies that are attached to and consistent with the same mathematical formalisms. Nevertheless the philosophy adopted is important and strongly colours all judgements. Hence, we shall call upon the meta-principles such as the theory of fresh eyes to assert a new choice of philosophy and to reveal what new concepts may be needed to quash the trouble with quantum mechanics.

Firstly let us query how we should explain quantum reality. This question can be answered in at least two ways that are fully compatible with the quantitative parts of quantum mechanics. In the Copenhagen interpretation, those (classical) quantities that are measured are considered the foundation of reality and it is not appropriate to ascribe any existence to any property that is not measured. Note that a real existence is also to be assigned to quantities that are measured indirectly. Thus in the traditional two slit experiment, it would be deemed to be invalid to ask which slit an electron went through if no measurement had been made to show which slit the electron transited. The quantum state is regarded as only a mathematical construct; it holds potential results of measurements and simply represents the embodiment of what we can possibly know about the system. Thus in the Copenhagen interpretation of quantum mechanics, the quantum state is not real; the quantum state is merely a useful mathematical construct.

However, familiarity with quantum mechanics leads to acceptance of the mathematics and, over time, some practitioners of quantum mechanics acquire a different view to that adopted in Copenhagen. In this alternate philosophy, the quantum state is taken to be a genuine representation of reality – after all the quantum state specifies *everything* that we could ever know about the system. Matter populating this state is describable as neither a particle nor a wave. Measurements constitute manipulations of the state into a form (eigenstate of measurement system) in which a classical variable has a well-definite value. This measurement eigenstate is not more real than the prior subsystem state. Hence, within this philosophical view, a description of the two-slit experiment maintains that the quantum state of the electron passes, in part, though both slits. Ascribing such a reality to the quantum state is clearly the simplest unprejudiced approach and thus this approach will be adopted here. This acceptance of the mathematical formalism follows the tradition[1] of believing what analyses of the natural world tells us about the universe. However, the assertion of a reality for the quantum state has dire consequences for causality (to be discussed in section 2.4).

2.3 Coordinate Representation of Nature

Let us now reflect on how we might want to represent the quantitative aspects of reality in a simple and consistent manner.

It is clear from many branches of physics that the basic quantitative mathematics required to abstractly mirror reality contains the mathematics of complex numbers rather than the mathematics of real numbers. This expression of complex mathematics is paramount in the principles of quantum mechanics but it is expunged in quantum measurement. Clearly a description of reality, that is consistently mathematical-complex yet is conceptually simple, would be appealing. Thus with justification from the theory of fresh eyes, this is a goal that shall be pursued.

Now, elementary mathematical representations have often given us profound insights that some have found hard to accept. The study of planetary motions showed that it was simpler to imagine that the Earth went around the Sun rather that the other way around. The study of relativistic transforms has shown that it is simpler to regard time as an imaginary fourth dimension than as a non-related entity. Thus these lessons suggest that we should fully heed the simple messages supplied by the way that mathematics describes reality.

Recall that the mathematical-imaginary number, $\hat{i}$, defined as the square root of minus one, has an ambiguous mathematical sign. For instance, if we look at the Fourier analysis equations, without loss of generality, we can consistently replace every $\hat{i}$ by its negative. Indeed, there is always a mathematical-sign ambiguity whenever we take a square root. In the mathematical field of complex analysis, when a function containing a square root is drawn on the mathematical-complex plane, this mathematical-sign ambiguity can require that the function be drawn on two planes that are joined across a branch cut.

If the mathematical-complex number is to be a building block of conceptual quantitative reality, then we should also accept that there is a physical reality behind this ambiguity in the mathematical sign of $\hat{i}$. Assigning two branches to the mathematical-complex space retains a capability to process this ambiguity. This dual-branch suggestion is counter to centuries of tradition during which we have drawn the basic mathematical-complex plane with just one branch. The argument for two branches in the space of numbers, is that the sign ambiguity in the mathematical-imaginary number, $\hat{i}$, is not an irrelevant accident and this ambiguity should not be treated as such. Mathematically we can and we should construct a two-branch structure that copes with both sign conventions.

Now, adopting a fully comprehensive mathematical-complex description of spatial reality would require that all space-time coordinates be made fully mathematical-complex and consequently space-time be given two branches. Thus we can ask what mathematical-real components are the missing terms that we should pair with the familiar position and time variables of 4-space.

Nature (through our physical principles) is forthcoming with strong evidence to answer this question. Firstly, in both Lagrangian and Hamiltonian formulations, classical mechanics uses (generalized) positions and momenta. These variables are the quantities that we need to know to be able to predict the evolution of an isolated system of particles. Secondly a position-momentum phase-space takes special significance in statistical mechanics defining the elemental volume that is uniformly accessed by equilibrium systems. Quantum mechanics subsumed these classical associations and added complementarity not between the position and momentum vectors but between their corresponding vector components. Thus our experience is that the vector components of position and momentum variables consistently pair in nature. This observation suggests that a mathematical-complex spatial coordinate should possess a mathematical-real part that is proportional to spatial position and a mathematical-imaginary part that is proportional to momentum.

It is necessary for the mathematical-real and mathematical-imaginary parts to have the same dimension. Thus let us define this generalized spatial coordinate as

$$\chi_\lambda = \hat{r}\, x_\lambda / s_U + \hat{i}\, (s_U / \hbar) p_\lambda \qquad \text{for } \lambda = 1, 2, 3 \qquad (2.1)$$

where x_λ and p_λ are components of the position and momentum vectors and s_U is a constant with dimensions of distance. The choice of $\hbar$ as the dimensioned constant to convert the mathematical-real and mathematical-imaginary components to related units, defines the scale for both s_U and χ_λ. For instance, we could have written this equation with Planck's constant, h, (or even another multiple) rather than Dirac's constant, $\hbar = h/2\pi$. However, this alternative form can be obtained from the former by transformations that turn s_U into $\sqrt{2\pi}\ s_U$ and χ_λ into $\sqrt{2\pi}\ \chi_\lambda$. Pedantically mathematical-real numbers will be denoted by the designator symbol, $\hat{r}$. This necessity for this symbol will only become relevant in part 2.

Now, the universe provides two natural size scales. On the small scale, general relativity and quantum mechanics combine to imply that the smallest sensible distance is of the order of magnitude given by the Planck distance, $\sqrt{G\hbar/c^3}$. (Here G is Newton's gravitational constant and c is the speed of light.) In contrast, the age of the universe, A_t, provides a very different large-scale distance, cA_t. However, the most practical value for s_U will be neither of these distance scales. Looking at the form of the equation for χ_i, we see that s_U plays the pivotal role of displaying whether distance and momentum values are either large or small. If s_U took on an extreme value, then all values would be permanently large or small. Thus we expect that s_U will take on an intermediate value. From another perspective, s_U provides a connection between our measurement standards (subjective view as to what is small or large) and universal units.

Hence if mathematical-complex algebra has a direct correspondence with the quantitative world we should attempt to formulate our more fundamental pictures of reality in terms of the dimensionless generalized spatial coordinate, χ_λ. Then, a six-dimensional statistical-phase-space integration becomes an integration over the complex planes of the dimensionless spatial components of χ

$$\int d^3x d^3p/\hbar \equiv \int d^{3^*}\chi \qquad (2.2)$$

In this complex space, elementary matter might be viewed as 'blobs' with the uncertainty principle requiring that a particle must maintain a minimum 'size' on each complex plane.

Note that the corresponding temporal dimension is

$$\chi_0 = \hat{r}\,(s_U/\hbar c)\,\varepsilon + \hat{i}\,(c/s_U)\,\tau \qquad (2.3)$$

where ε and τ represent the energy and time. Thus nature does seem to be suggesting that the relevance of complex mathematics could be fundamental and extensive.

2.4 Time Sense

Now, the laws of physics are time-direction invariant (or more precisely CPT invariant). Classically we have understood this CPT-invariance symmetry to mean that if we were to flip the time direction (T), the spatial parity (P) and charge conjugation (C) throughout the universe, then history would reversibly unfold backwards. The quantum rules of measurement are an exception to this strict classical deterministic form of behaviour. In the Copenhagen interpretation, quantum measurement is supposedly non-deterministic; hence theoretically reversing the flow of time after an experiment will not necessarily lead to a recovery of the original state. Measurement changes the quantum state forward in future time but measurement does not change the quantum state backwards in past time. Thus the interpretation of the measurement rules can appear to hold a strong time-sense bias.

This sense of time directionality in the interpretation of quantum measurement is associated with the notion of causality. Note that there are two forms of description for the principle of causality. In the first weak form, cause and effect are simply reversibly correlated quantities that are not being used to convey information. An example of the weak form of causality would be the correlated changes in the pressure and volume of an isothermal gas. When changes occur, causality may be used to attribute which 'action' is the cause and which 'consequence' is the effect. Reversing time would see the exchange of action and consequence but with no change in the governing laws.

The strong form of the description of causality is one that says that we may transmit information into the true future (ie within the forward time cone) only. Consider information written by pencil on paper on one Monday morning and rubbed off the following Monday. The time asymmetry doesn't come from whether the information was deemed to be written by the pencil and travelled forward in time from the first Monday or from whether the information was deemed to be 'written' by the eraser and travelled backward in time from the second Monday. The asymmetry comes from the meaningful application of the pencil on the first Monday and the destructive intent in the erasure the following Monday. Thus there is some danger that this second form of causality may simply be a manifestation of our perspective (as unremitting forward-time travellers) rather than be a useful revelation of any genuine tendency of nature.

It is possible to organize a measurement for which this second strong assertion of causality fails. When a measurement is made, the Copenhagen interpretation requires that the quantum state instantaneously change. This state change must occur even when the measured subsystem is widely dispersed spatially. Thus it is possible to organize and even to demonstrate[2], consequences of a measurement that effectively propagate outside of the forward light cone. For the Copenhagen interpretation, the quantum state is not a real entity and thus the quantum state does not have to satisfy causality. The correlated measurements that are made inside or outside the forward light cone, demonstrate the weak-causality principle. On the other hand, if we attribute a reality to the quantum state, then we must at the very least abandon determinism. Then, our laws may be CPT symmetric yet the outcomes of a reversal of symmetries could be uncertain.

Quantum mechanics has been triumphantly successful in describing so much of reality. The central entity in a quantum description is the quantum state. Challenges to a quantum description of reality that attempt to restore classical deterministic notions of reality have been experimentally disproved. It seems only reasonable to presume that the quantum state is part of material reality. However, this attribution of reality to the quantum state necessarily requires that we accept some form of acausal non-deterministic change within reality.

In the light of these discussions, we shall attempt to form a partial illustrative model of reality that resembles quantum mechanics yet the model will not be beholden to unnecessary classical ideology. This model will be referred to as free mechanics; the model will be fully time-direction symmetric and admit acausal interactions. Now, we have already proposed that the spatial coordinate be mathematical-complex and that there be two branches to the space-time continuum, each branch with a different time-direction sense. Note that this two-branch structure is fully consistent with all physical laws and with the principles of relativity. The next section discusses how the two branches of space-time can accommodate acausality during measurement.

2.5 Space-Time Anti-Discrimination

In the equations of physics, time behaves as an additional dimension thus time is rightly regarded as such. The assignment of time as an additional spatial dimension is clearly demonstrated by the coordinate transformations, between frames of reference in the theory of special relativity, that mix what was previously regarded as 3-space and time. If we accept this assignment as revealing a true equivalence of time with a fourth dimension, then perhaps we should accept that any physical process that can be displayed in space should have an equivalent process displayed within time.

Now, according to quantum mechanical principles, the interaction of a particle state with a one-dimensional spatially localized (classical-interaction) potential *causes* some of the state to be reflected or scattered backwards in 3-space and some of the state to be scattered forward in 3-space. Hence, if there is a genuine symmetry linking 3-space and time, we might expect[3] that a particle state interacting with a *temporally* localized (classical) potential could *cause* some of the state to be scattered backwards in time and some of the state to be scattered forwards in time. Note that we are chaining these processes causally not chaining them temporally.

In most measurements, the state of the system being measured undergoes a rapid change to become an eigenstate of a traditional classical variable. Thus we expect that such a measurement process will cause a strong reflection of matter backwards in time (ie onto any backwards-in-time branches of space). Essentially, at the point of the measurement, the branches become united with the state of the classical variable.

At a stroke, this assertion (of comparable treatment of 3-space and time) breaks strict causality, initiates indeterminism, and enables us to display measurement law as time-direction symmetric. By giving equal status to both time-direction branches, a backwards-in-time propagating state will itself necessarily be reflected forwards in time at an earlier backwards-measurement (the experimental set-up that defined the initial condition of the subsystem). These effects generate endless logistic loops between pairs of consecutively linked measurement systems. Furthermore, without violating any relativistic principles, disconnected parts of a subsystem will be able to communicate without recourse to faster-than-light interactions, via accessible common earlier-in-time measurements. For some unprepared systems of matter it is possible that the appropriate set-up event could even be the big bang.

Hereby, we have overcome major objections to the state becoming a real entity (these objections claim that measurement entails interactions incompatible with relativistic principles); we have enabled measurement and propagation processes to be conceptually reunited (measurement is like applying a boundary condition in time and can be replaced by propagation if the time dependence of the interaction is resolved) and we have given meaning to the process of measurement (the transfer/union of the data of quantum states across the different time-sense branches). Measurement is interpreted as no more than propagation accompanied by a strongly time-dependent interaction. The results of this measurement are determined by the associated propagation of the states both forward and backward in time.

Here we have attempted to give new meaning to measurement. The next section will give more meaning to the quantum state.

2.6 Over-Simplified Model of Free Mechanics

It is likely that wherever the differences between the principles of quantum mechanics and the rules of measurement (as listed in table 2.1) are irreconcilable, then we should probably favour the least restrictive characteristic. This selection is consistent with the way in which we have favoured mathematical-complex numbers over mathematical-real numbers and acausality over causality. Furthermore a less restrictive interpretation leaves less to be justified. In discussing these characteristic differences, however, we have suggested that consistency requires further unorthodox features in a modified approach that is able to reconcile the principles and rules of quantum mechanics. These unorthodox features include a double-branch structure for all the mathematical-complex numbers. A double-branch structure for space provides two branches with different time-direction sense.

We are now in a position to put some of these ideas together. We shall attempt to do this in a relatively *ab-initio* but casual manner. The presented model is not a serious attempt to build a description of reality. The model is a highly restricted sketch intended to support the contention that our selected attributes may be what are needed to fix the inherent incompatibilities residing within quantum mechanics.

We choose to allow substantial freedom to our system. Thus following the approach of Maxwell and Boltzmann's statistical mechanics, let us consider a statistical analysis for a massless non-interacting instance of a multi-particle field of the universe. This multi-particle field will comprise many spatial coordinates each representing an instance of a particle, labelled by index i (and with a particular generalized coordinate $\chi^{(i)}$).

We shall presume that the condition of the field (used to define an ensemble) is appropriately described by the values of those physical quantities that are universally conserved. These universal invariants are those quantities that are recognized as being conserved for interactions between instances of the field, namely the *cumulative*[4] generalized coordinate itself. Now, this macro-state can be constructed in different ways from individual choices for the values of generalized coordinates, $\chi^{(i)}$, with temporal fluctuations taking the universe quickly through many of these choices with equal likelihood. There are an astronomically large number of such instances and hence the number of arrangement choices, Ω, available to the system is both exceedingly large and varies greatly with the particle's coordinates. To obtain a slowly varying (mathematical-complex) value of experiential order of magnitude, it is necessary to work with the entropy-like logarithm of the number of choices, $\ln\Omega$. Recognizing that the cumulative generalized coordinate is a relative vector we explicitly display this dependence

$$\ln\Omega = \ln\Omega\left(\Xi - \Xi_0\right) \qquad (2.4)$$

where $\Xi = \sum_i \chi^{(i)}$ and Ξ_0 is ensemble average cumulative coordinate.

Furthermore, there can be no preferred direction in the universe; thus the number of accessible states is dependent upon powers of the scalar product, $a = (\Xi - \Xi_0)^2$.

Starting from $a = 0$ in a McClaren expansion, consider a small fluctuation of one isolated particle instance, i, leaving the other instances unchanged.

Then, assuming $\ln\Omega$ has an analytic dependency on generalized coordinates, one can write

$$\ln\Omega = \ln\Omega_0 + \frac{1}{2}\beta\frac{\partial a}{\partial\chi_\mu^{(i)}}(\chi - \chi_0)_\mu^{(i)} + \ldots \tag{2.5}$$

where

$$\beta = 2\left(\frac{\partial \ln\Omega}{\partial a}\right)_0 \tag{2.6}$$

plays the role of the inverse of a statistical temperature from which the randomness of the field may be described.

The difference, in the number of states available on changing just one instance, is very much smaller than the total number available. Thus we expect this series (equation (2.5)) to be rapidly convergent. Hence dropping the higher-order terms, the number of available choices has a Gaussian appearance

$$\Omega \propto \exp\left(\beta(\chi^{(i)} - \chi_0)^2\right) \tag{2.7}$$

as appropriate for a random continuous independent variable. The mathematical-complex number of choices, Ω, quantifies the (mathematical-complex) likelihood of the universe being in this state. Consequently the probability of this isolated particle being at 4-position $(\underline{x},\tau)$ with 4-momentum $(\underline{p},\varepsilon)$ is simply

$$\begin{aligned} P(\underline{x},\tau;\underline{p},\varepsilon) \propto \exp\Big\{&\beta\left(\underline{x}^2 - c^2\tau^2\right)/s^2 + i\ 2\beta(\underline{p}\underline{x} + \varepsilon\tau)/\hbar \\ &+ (\varepsilon^2 - \underline{p}^2c^2)s_U^2/\hbar^2c^2\Big\}. \end{aligned} \tag{2.8}$$

For the particle to be completely non-interacting, it needs to be massless and travel at light speed. Then, the (mathematical-complex) probability distribution will take the familiar form for a free-space wave-function (i.e. the spatial representation of a free single-particle quantum state)

$$P(\underline{x},t;\underline{p},\varepsilon) \propto \exp\left\{-i\ (\underline{p}\underline{x} + \varepsilon t)/\hbar\right\} \tag{2.9}$$

where I have taken $\beta = -1$. This result provides a physical basis for understanding the quantum state and the quantum rules of measurement. Furthermore, the negative[5] value for β suggests that nature actually prefers the 4-momentum vector to be the complex-real component of the coordinate with the 4-position vector as its complex-imaginary complement. This may seem counterintuitive for some but motion is inherent in everything. The fundamental importance of motion is understood in special relativity. It is easy to overlook that everything in the universe is in a state of flux. Matter, which is spatially 'at rest' in a particular reference frame, is moving at maximal speed through time. In contrast, time actually stops flowing for matter that is moving at maximal speed through space. Indeed whatever the velocity, according to special relativity, everything can be said to move at precisely the same four-vector speed – physical motion is universal.

Thus in this analysis, the wavefunction is treated as a true mathematical-complex probability, not a probability amplitude. When we make a measurement of a classical variable, the outcome is determined by the mathematical-complex probability on both time-sense branches. Hence, we calculate the result obtained in a measurement by multiplying the probabilities of achieving this result on both time-direction branches. This multiplication explains the form of the expression that we use to compute measurement outcome probability.

This model holds the implicit suggestion that quantum mechanics may be a form of statistical mechanics expressed using a suitable mathematics and space.

2.7 Conclusion

The proposition presented here is that we do not require three long-held notions:

1. the notion that reality be represented by mathematical-real numbers
2. the notion that measurement rules could be time-direction asymmetric
3. the notion that the past is fixed and that the future is fluid.

Instead we should accept the following new notions:

1. the notion that reality be represented by mathematical-complex numbers
2. the notion that measurement rules could be time-direction symmetric
3. the notion that parts of both the past and future are fluid.

The benefits of accepting this approach are mainly philosophical. The new approach attributes meaning to the state wavefunction as a complex distribution function on a branch of the space-time continuum; the approach explains the natural role of phase space in statistical mechanics as a equivalent representation of a real mathematical-complex coordinate space; the approach provides a physical interpretation of a measurement result as the combined probability of the value on two mathematical branches of space-time; the approach describes the measurement process in a relativistically appropriate manner; the approach suggests that we can derive our ordered laws of natures from an underlying description of reality inherently based upon random processes; the approach banishes the demons of determinism; the approach places time and space on a more equal footing with time-variation potentials giving rise to reflections in time just as space-variation potentials give rise to reflections in space; the approach combines propagation theory with measurement theory allowing partial or full measurements depending on the extent to which the branches are forced to hold the same result.

However, what we have presented to date is not a complete model. At best, the arguments apply only to an isolated non-interacting particle. The results provide no new testable hypothesis. No prediction of the model formulation can compete with the predictions of quantum mechanics. However, the proposed model is akin to Planck's quantum hypothesis. The arguments only reinterpret weird aspects of what is already known.

Nevertheless we have completed the instruction embodied in the second step of the map. The question that we have answered is quite specific. To progress this step we have asked what is it that we would need to *add* to the existing quantum mechanics formulation that would resolve quantum mechanics' difficulties. These difficulties are philosophical; we have come up with revised interpretations. We do not need to answer what we need to give up. Essentially we have come up with two additions. The first addition is the widespread role of mathematical-complex numbers on branches of space; the second addition is an underlying random vision of reality.

The first suggestion is philosophically attractive; there should be no special place for mathematical-real numbers. This suggestion will become the subject of the generalizations involved in the next step. The second suggestion is appealing from one viewpoint; the origin of chaos, of randomness and of disorder needs no imposed justification. However, from another viewpoint this proposition is disconcerting, since the proposition seems to source the reality of our experience out of nihilism. An investigation of the role of randomness will be given later where it has been considered both as a dualism candidate and as part of a trend in scientific progress. These later discussions will satisfactorily banish nihilism.

Notes

[1] The Earth goes around the Sun. Time is a fourth dimension. The correlated missing energy and momentum in beta-decay are carried away by a massless neutrino.

[2] In the traditional EPR experiment, a two-particle entangled state is created and the particles (perhaps electrons, perhaps photons) are allowed to travel a long distance away from each other. According to quantum mechanics a measurement, on one particle, of the entangled property is instantaneously transmitted to the other ie faster than light speed. This effect is indirectly tested experimentally using Bell's theorem. These experiments confirm that measurements on the two particles are correlated but the correlation is not independent of separate choices that we can make at the time of individual measurements.

[3] Once again we are appealing to the theory of fresh eyes. We are taking off the blinkers provided by the standard paradigm and asking if there is a simpler way of describing what we know actually occurs.

[4] Here we are referring to the sum of the individual mathematical-complex coordinates. The sum of momenta gives the conserved total momentum of the system. The system's centre of mass is calculated from the sum of positions.

[5] A negative value is needed for equation (2.7) to take the Gaussian form and for the system to be stable.

Part 1

Chapter 3 : The Master Principle

3.1 Introduction

Every description of reality has its master principle. The master principle colours the worldview. The master principle of classical mechanics maintains that everything is made from discrete point particles interacting via forces at a distance. The master principle of quantum mechanics imagines a world entirely built from quantized field elements with local interactions. In a similar manner, the master principle of the (yet to be exposed) new reality will indicate a distinctive core value of the new paradigm.

Step 3 of the map is where we identify and explore the new master principle that will pervade our new description of reality. Previous master principles have started from humble origins. For instance, Planck proposed discreteness in the energy absorbed and emitted by atoms. The expansion of this minor principle into a master principle created the dominant concept that shattered the mould of (laboratory-scale) classical mechanics. Thus the instruction in the third step is to take the chosen output from step two of our map (ie a new addition to the description of reality) and to identify a master principle that encompasses this addition.

In step 2, we identified two necessary additions to quantum descriptions: a complete lack of constraint and the applicability of universal mathematical-complex quantities. As an absolute ideal, complete disorder should be regarded as a concept of last resort, ie an absolute is expected in an ultimate description of reality. It is unprofitable to be too ambitious. The natural generalization into a master principle would involve requiring everything becoming an absolute and thus the generalization would require that our paradigm shift take us to the ultimate theory of reality. Such a choice immediately invites questions as to how anything can be justified when nothing is imposed. At this stage of our arguments, this question is too imprecise to be tackled. Therefore we shall be less adventurous and we will turn to the alternatively offered addition – the adoption of mathematical-complex quantities. Interesting this approach will be shown to generate an intermediate step from which the ultimate fundamental world of absolutes is visible. Hence, in section 13.4.10, we shall answer the fabulous conundrum: how nothing can explain everything.

We are to investigate how to make more general a formulation of science based upon mathematical-complex numbers. Let us start by reconsidering what we know about the mathematical-complex pairing of observable quantities in current science.

3.2 The Pairing of Observables

We have argued that the components of mathematical-complex numbers are paired complementary conjugate variables. Thus we anticipated that we will construct a mathematical-complex variable from position and momentum variables. However, let us consider in more detail the identification of pairs of conjugate variables.

Quantum mechanically complementary variables are those variables that do not commute and hence such variables cannot be known simultaneously[1]. Note that written in general form Heisenberg's uncertainty principle (quantifying this restriction on simultaneous knowledge of observables) is

$$\Delta p_\lambda \Delta q_\lambda \geq \tfrac{1}{2}\left|\left\langle \left[p_\lambda, q_\lambda \right]_- \right\rangle\right| \tag{3.1}$$

The brackets refer to evaluating the commutator for the pair of operators, taking an expectation value and then taking the mathematical-complex modulus.

Clearly if the pair of variables (p_λ, q_λ) is quantum complementary, then so are many functions of these variables including a simple scaling such as $(p_\lambda / m_0, q_\lambda)$ where m_0 is a rest mass. Indeed the mismatched units accredited to p_λ and q_λ suggest that we may be able to construct a more enlightening pair of matched-unit conjugate variables.

Now, most quantum principles have a classical analogue. In classical mechanics we may formally define generalized position and momentum variables using so-called canonical definitions. Recall that the Lagrangian approach is formulated in generalized position, q_λ, and velocity, $\dot{q}_\lambda$, variables. The Hamiltonian approach is formulated in generalized position, q_λ, and momentum,

$$p_\lambda = \frac{\partial L}{\partial \dot{q}_\lambda} \tag{3.2}$$

variables. However, each approach can be defined in terms of the other approach. Thus we also have

$$H = \sum_\lambda p_\lambda \dot{q}_\lambda - L, \; \dot{q}_\lambda = \frac{\partial H}{\partial p_\lambda} \text{ and } \dot{p}_\lambda = -\frac{\partial H}{\partial q_\lambda}. \tag{3.3}$$

Furthermore we see the pairing of q_λ and p_λ in the Poisson bracket,

$$\left[X, Y\right]_{Poisson} = \sum_\lambda \left(\frac{\partial X}{\partial q_\lambda}\frac{\partial Y}{\partial p_\lambda} - \frac{\partial X}{\partial p_\lambda}\frac{\partial Y}{\partial q_\lambda} \right) \tag{3.4}$$

which acts like the commutator of quantum mechanics. Hence with $Y = H$ the Poisson bracket identifies the classical constants of motion.

These classical differential expressions (eg equation (3.3)) lead to the operator representations of quantum mechanics eg for momentum and energy

$$p_\lambda = -\hat{i}\hbar \frac{\partial}{\partial q_\lambda}, \quad H = \hat{i}\hbar \frac{\partial}{\partial t}. \tag{3.5}$$

Then the commutator becomes

$$[p_\lambda, q_\lambda]_- = -\hat{i}\hbar. \tag{3.6}$$

Furthermore we get the Fourier-conjugate pairing appearing in the transformation exponential

$$< q_\lambda | \psi >= \int dp_\lambda \exp(\hat{i} p_\lambda q_\lambda / \hbar) < p_\lambda | \psi >. \tag{3.7}$$

Once again there is no necessary sense of scale associated with the concept of complementarity. The constant[2], $\hbar$, is introduced to satisfy the empirical energy-frequency ($E-\nu$) correlation:

$$E = h\nu \tag{3.8}$$

recognized by Planck. Here h is Planck's constant and $\hbar = h/2\pi$. Hence there is freedom in setting up the conjugate pairing of variables. We may even change the units of the conjugate variable provide that we make the appropriate change to the corresponding units of phase-space volume.

3.3 Hyper-Complex Numbers

We have observed that mathematical-complex numbers are an essential part of our physics. Indeed mathematical-complex numbers can be seen to be relevant from two main standpoints. Firstly these numbers arise naturally in a mathematical-development sequence that passes from natural numbers through rational numbers and mathematical-real numbers. Thus we observe that the system of mathematical-complex numbers obeys the generally accepted axiomatic laws[3] that define integral-domain numbers (ie the commutative, associative, distributive, identity and cancellation laws). Secondly, whereas classical mechanics was formulated with only mathematical-real numbers, we found that it was more natural to formulate quantum mechanics with mathematical-complex numbers. Clearly classical mechanics could have been formulated with mathematical-complex numbers. However, there appears to be no significant benefit in doing this. Nevertheless classical mechanics does contain many pairings of physical quantities. This pairing effectively occurs in both the Hamiltonian and Lagrangian formulations of classical mechanics, the pairing also appears in the formulation of classical statistical mechanics and even in Heisenberg's uncertainty principal, which although an essential ingredient of quantum mechanics, had its origins in classical measurement theory. Thus mathematical-complex numbers do appear to be a natural starting point for a scientific theory of reality.

However, other equally important number forms are engrained in Physics. One of these natural number forms is the 3-vector form. Vector arithmetic can be built as a construct of conventional scalar number theory. Thus whereas we can ask: *What is the square root of minus one?* - one does not normally ask the question: *What is the square root of a 3-vector?* We don't ask such a seemingly ridiculous question because the square (ie vector or cross product with itself) of a finite 3-vector is always zero. However, the manipulation of special-relativity equations routinely involves the transformation of vectors. Thus when Dirac derived the first form for relativistic quantum mechanics by effectively taking the square root of the D'Alembertian[4] operator, he found that (special) relativistic quantum mechanics required further number components that are a non-commutative under the multiplication operation. Although a matrix representation is generally used to incorporate the loss of multiplicative commutation, these new components are closely related to the quaternions numbers conceived by Hamilton.

A quaternion number (or quaternion) is comprised of a scalar and 3-vector component with a special feature that the square of a 3-vector component is a quaternion-scalar. As algebraic numbers (rather than posited composite constructs), quaternions can potentially explain the presence of 3-vectors in the physical world. Furthermore, quaternions have a natural association with mathematical-complex numbers, as they were devised as a higher-dimensioned generalization of mathematical-complex numbers.

3.3.1 Quaternion Essentials

Quaternions will prove to be a central feature of our new hypothesis. Hence an introductory summary of quaternion numbers is provided for those readers who are less familiar with quaternions.

Recall that numbers are simply the elements of an algebra. We may say that an algebra is a set of compatible rules for manipulating its elements (which are its numbers) ie an algebra describes the mathematical operations that the algebra's numbers can perform. The algebraic set of rules can be reduced to a minimum-sized set of independent rules. The size of this minimum set varies with different style algebrae. There are no necessary axioms. Further, for this minimum set of axioms, there may be several different choices that correspond to the same algebra. Hence, as there are many algebrae, the algebra that appropriately describes the quantitative aspects of the universe is decided by science and not decidable by mathematics.

In 1835 Sir William Rowan Hamilton recognized the geometrical significance of describing mathematical-complex numbers as pairs of mathematical-real numbers on the complex plane.

$$\underline{\underline{z}} = x\hat{r} + y\hat{i} \equiv (x, y) \qquad (3.9)$$

x and y are mathematical-real numbers. $\hat{r}$ and $\hat{i}$ are designators labelling the mathematical-real and mathematical-imaginary component axes.

Table 3.1

Designators $\pm\hat{r}$ and $\pm\hat{i}$ form a group. The group's product rule requires that $\hat{r}$ act as an identity for the group.

	$\hat{r}$	$\hat{i}$
$\hat{r}$	$\hat{r}$	$\hat{i}$
$\hat{i}$	$\hat{i}$	$-\hat{r}$

Thereafter Hamilton began to think about a generalization that would be able to describe the geometry of space (presumed at the time to be 3-dimensional). Hamilton achieved this goal in 1843 by defining quaternions numbers as a pair of a mathematical-real number and 3-vector of mathematical-real numbers.

$$\underline{\underline{\Gamma}} = a\hat{\gamma}_0 + u_1\hat{\gamma}_1 + u_2\hat{\gamma}_2 + u_3\hat{\gamma}_3 \equiv (a, \underline{u}) \equiv (a, (u_1, u_2, u_3)) \qquad (3.10)$$

Table 3.2

The product rule for quaternion designators requires that $\hat{\gamma}_0$ acts as an identity element and requires that $\hat{\gamma}_1$, $\hat{\gamma}_2$ and $\hat{\gamma}_3$ act as Cartesian vector components which like the $\hat{i}$ designator have a square that is the negative of the algebra's identity.

	$\hat{\gamma}_0$	$\hat{\gamma}_1$	$\hat{\gamma}_2$	$\hat{\gamma}_3$
$\hat{\gamma}_0$	$\hat{\gamma}_0$	$\hat{\gamma}_1$	$\hat{\gamma}_2$	$\hat{\gamma}_3$
$\hat{\gamma}_1$	$\hat{\gamma}_1$	$-\hat{\gamma}_0$	$\hat{\gamma}_3$	$-\hat{\gamma}_2$
$\hat{\gamma}_2$	$\hat{\gamma}_2$	$-\hat{\gamma}_3$	$-\hat{\gamma}_0$	$\hat{\gamma}_1$
$\hat{\gamma}_3$	$\hat{\gamma}_3$	$\hat{\gamma}_2$	$-\hat{\gamma}_1$	$-\hat{\gamma}_0$

a, u_1, u_2 and u_3 are mathematical-real numbers and individually the mathematical-real number pairs (a,u_1), (a,u_2) and (a,u_3) resemble mathematical-complex numbers. $\hat{\gamma}_0$, $\hat{\gamma}_1$, $\hat{\gamma}_2$ and $\hat{\gamma}_3$ are designators labelling the components of the quaternion.

Hamilton strongly publicized quaternions. However, physicists found vector arithmetic more useable after Josiah Willard Gibbs formalized vector-product notation by introducing the scalar and vector products.

The product of two quaternions can be written in terms of scalar and vector products as

$$(a_1,\underline{u}_1)\times(a_2,\underline{u}_2)=(a_1a_2-\underline{u}_1.\underline{u}_2,a_1\underline{u}_2+a_2\underline{u}_1+\underline{u}_1\times\underline{u}_2). \qquad (3.11)$$

3-vectors are seen to be incomplete forms of a quaternion. Although vector arithmetic provides two styles of multiplication, quaternions only have one analytic multiplication operation. Now, the square length of a quaternion is obtained by multiplying a quaternion by its quaternion conjugate denoted by a raised dagger, †.

$$(a,\underline{u})^\dagger\times(a,\underline{u})=(a,-\underline{u})\times(a,\underline{u})=(aa+\underline{u}.\underline{u},0) \qquad (3.12)$$

Taking a quaternion conjugate is a non-analytic operation; the operation is performed on the identified components it is not an algebraic operation upon the entire quaternion number. More disturbingly some people generalize this square length result to provide a contrived non-analytic scalar product of quaternions

$$\begin{aligned}(a_1,\underline{u}_1)\cdot(a_2,\underline{u}_2)&=\frac{1}{2}\left((a_1,\underline{u}_1)^\dagger\times(a_2,\underline{u}_2)+(a_2,\underline{u}_2)^\dagger\times(a_1,\underline{u}_1)\right)\\&=(a_1a_2+\underline{u}_1.\underline{u}_2,0).\end{aligned} \qquad (3.13)$$

Consideration of the requirement for analyticity is instructive. Sometimes we might accept non-analytic operations when we are changing our conceptual framework. For instance, the involvement of mathematical-complex conjugates in quantum measurement can be attributed to the stark mismatch between our quantum and classical pictures of reality. In a quantum measurement we are required to convert information in the quantum wavefunction into a classically real value. Classical reality is built on very different precepts from those of quantum reality. However, if we investigate quantum mechanically the limiting process in which aspects of quantum reality approach classical ideals then our analysis of these aspects should remain analytic.

Furthermore when we are not changing the conceptual picture of reality we expect our formalisms to be analytic. Consequently if we need to resort to non-analytic operations then we cannot be using the most appropriate algebra. If the decree of mathematics applies then there will be a set of algebraic axioms and a corresponding system of numbers that is suited to this conceptual stratum (ie storey of the edifice) of reality. Thus analyticity tells us about the suitability of our mathematical formalism.

Finally let us note how we may represent the quaternion designators using 2 x 2 matrices made with mathematical-complex components

$$\hat{\gamma}_0 \equiv \begin{pmatrix} 1 & 0 \\ 0 & 1 \end{pmatrix} \quad \hat{\gamma}_1 \equiv \begin{pmatrix} i & 0 \\ 0 & -i \end{pmatrix} \quad \hat{\gamma}_2 \equiv \begin{pmatrix} 0 & 1 \\ -1 & 0 \end{pmatrix} \quad \hat{\gamma}_3 \equiv \begin{pmatrix} 0 & i \\ i & 0 \end{pmatrix} \qquad (3.14)$$

or 4 x 4 matrices made with mathematical-real components

$$\hat{\gamma}_0 = \begin{pmatrix} 1 & 0 & 0 & 0 \\ 0 & 1 & 0 & 0 \\ 0 & 0 & 1 & 0 \\ 0 & 0 & 0 & 1 \end{pmatrix} \quad \hat{\gamma}_1 = \begin{pmatrix} 0 & -1 & 0 & 0 \\ 1 & 0 & 0 & 0 \\ 0 & 0 & 0 & 1 \\ 0 & 0 & -1 & 0 \end{pmatrix}$$

$$\hat{\gamma}_2 = \begin{pmatrix} 0 & 0 & 1 & 0 \\ 0 & 0 & 0 & 1 \\ -1 & 0 & 0 & 0 \\ 0 & -1 & 0 & 0 \end{pmatrix} \quad \hat{\gamma}_3 = \begin{pmatrix} 0 & 0 & 0 & -1 \\ 0 & 0 & 1 & 0 \\ 0 & -1 & 0 & 0 \\ 1 & 0 & 0 & 0 \end{pmatrix}. \qquad (3.15)$$

These matrices, representing the quaternion designators, have unit determinant. The matrices corresponding to the quaternion-vector designators have zero trace; the trace of the identity (quaternion-scalar) designator is the appropriate dimensionality. The conjugate operation for quaternions switches the sign of an odd number of designators ie reverses the spatial parity of the described vector space. For the mathematical-real representation, the transpose operator (which switches columns and rows) is equivalent to the quaternion conjugate operator (ie flips the quaternion-vector sign) and is equivalent to the inverse operator. The mathematical-complex representation of the designators requires an additional mathematical-complex conjugation.

The quaternion algebra displays several other geometric symmetries. For instance any cyclic rotation of the quaternion-vector designators eg

$$\begin{bmatrix} \gamma_1 \\ \gamma_2 \\ \gamma_3 \end{bmatrix} \rightarrow \begin{bmatrix} \gamma_2 \\ \gamma_3 \\ \gamma_1 \end{bmatrix} \qquad (3.16)$$

leaves the algebra, expressed in the relational table 3.2, invariant and likewise a reversal of the cyclic order of the quaternion-vector designators accompanied by their negation also preserved the algebraic relationships.

$$\begin{bmatrix}\gamma_1\\ \gamma_2\\ \gamma_3\end{bmatrix} \to \begin{bmatrix}-\gamma_3\\ -\gamma_2\\ -\gamma_1\end{bmatrix} \qquad (3.17)$$

There are also further symmetries of the type

$$\begin{bmatrix}\gamma_1\\ \gamma_2\\ \gamma_3\end{bmatrix} \to \begin{bmatrix}-\gamma_1\\ -\gamma_2\\ \gamma_3\end{bmatrix}. \qquad (3.18)$$

3.3.2 A Generalized Picture

Mathematical-complex numbers do appear to be required to describe some genuine aspects of reality. Likewise, quaternions (as a formal way to represent 3-vectors) could be said to be required to describe other aspects of reality (provided the spatial universe is $2\times(1+3)$ dimensional). Quaternions fit within a hierarchy of number types[5]. Mathematical-real numbers are distinguished by their ordered sequence capability – they will fill an unbounded one-dimensional line of mathematical-real numbers. Mathematical-complex numbers lose this strict ordering property but they satisfy the remainder of the axiomatic arithmetic rules. These numbers will fill an unbounded two-dimensional plane of numbers. Quaternions are a further generalization with the loss of commutation under the product operator. Quaternion numbers will fill an unbounded four-dimensional volume of numbers. If we need a formulation of reality based upon mathematical-complex numbers then perhaps we also need a formulation based upon quaternions. Hence, we have arrived at a principle that can be given a fruitful universal applicability – a worthy master principle. This principle requires that the natural algebra involves numbers that are more generalized or less specialized than mathematical-real numbers.

The quaternion-vector designators are a generalization of the imaginary designator, $\hat{i}$. Thus, just as there is a mathematical-sign ambiguity with mathematical-imaginary designator, $\hat{i}$, there is also a sign ambiguity with the vector part of a quaternion[6]. If this ambiguity exists in the mathematics and this mathematics describes reality then it is reasonable to argue that the sign ambiguity does exist as a physical reality and argue that we should require further pairs of branches of reality - in this case with different spatial parity. We have already argued that there should be two branches of 8-space with different directions of temporal evolution which are interchanged by the operation of mathematical-complex conjugation. Now, we make exactly the same argument when we consider the quaternion conjugation operation. A full correspondence of the mathematics with physical reality requires more branches that are related by taking the quaternion conjugate.

Hence we arrive at a description of 8-space with four branches differentiated by combinations of the direction of time flow and their spatial parity. Everything that we observe (in measurement) about our world is to be a superposition of the features that exist on each of these four branches[7] of 8-space. Geometrically we can link these branches with branch cuts, one cut on the mathematical-real axis corresponding to the mathematical-complex arbitrariness and one cut on the quaternion-scalar axis to corresponding to the quaternion arbitrariness.

3.4 Conclusion

The aim of these early steps of the map is to identify features that will build a new picture of reality. Thus the aim is not yet to construct a rigorous formalism.

We have identified a master principle. The master principle says that we should form all basic quantitative descriptions from less-specialized numbers (ie less specialized than mathematical-reals). We should not start with very specialized numbers such as the mathematical-real numbers or scalar numbers. In step 2, we accepted the need for mathematical-complex numbers. With the request for less specialized numbers, we don't necessarily have to accept quaternions. With a suitable justification, we could accept some other algebra that can introduce vectors in a natural manner. Nevertheless fundamental quantitative reality will not be written in terms of scalars, vectors or tensors, except as components or representations of some more appropriate number form. Actually, in part 2, we shall observe that quaternions are an important fundamental number form.

In the next chapter we shall investigate the impediments that confuse the construction of the anticipated synthesis.

Notes

[1] Non-commuting variables may not have common eigenvectors.

[2] $\hbar$ (h-cross or h-bar) is sometimes known as Dirac's constant.

[3] These axiomatic laws are discussed in sections 9.2.2 and 10.2.2.

[4] The D'Alembertian operator, $\frac{1}{c^2}\frac{\partial^2}{\partial t^2}-\frac{\partial^2}{\partial x^2}-\frac{\partial^2}{\partial y^2}-\frac{\partial^2}{\partial z^2}$, is the 4-space equivalent of the Laplacian.

[5] The axiomatic basis for this sequence of algebrae is explained in table 10.2.

---

[6] All quaternion-scalar results are formed from summations of even powers of quaternion-vectors. Flipping the sign of all quaternion-vector components (ie taking the quaternion conjugate) gives us an equivalent parity mirror world.

[7] Eventually this picture will be expanded to reveal eight branches along with the eight dimensions revealed in chapter 2.

Part 1

Chapter 4 : Dualities

4.1 Introduction

As our theories gradually evolve and become better able to describe nature, most fundamental concepts will carry over from an older version of a theory to its improved replacement. However, major evolutionary steps require the development of new concepts that subsume a range of older concepts. Currently our representations of reality have too many concepts for our theories to be taken to be close to an ultimate explanation of reality. We anticipate that when we get close to a foundation we will observe that, with each paradigm shift, the total number of concepts should steadily decrease. Within the context of the preceding theory, the concepts to be displaced will be quite distinct. For instance, quantum mechanics was able to take particle-like behaviour and wave-like behaviour and to incorporate these behaviours into the concept of a quantum state. To some extent, one can view second quantization as merging specification of actions (as previously described by operators) with objects (as previously described by states).

Important dualities are characterized by a strong incompatibility (reluctance to merge) within the context of the prevailing worldview. Through this clash, dualities identify the impediments that are forcing scientific progress to resort to a total reformulation. Thus dualities provide important clues as to how to clothe the new theory with new concepts. Now, in the preceding theory that is to be replaced, one side of the duality will be dominant and one side will be subordinate. However, we will fabricate the union of these duelling concepts by reversing this dominance. For example, by starting with a formulation based upon the classically subordinate wave concept and applying the master quantum principle, Schrödinger showed how to unify both particle and wave behaviours into a single new entity: the quantum state.

Some of the earliest recorded duality dilemmas were those considered by the pre-Socratic philosophers[1]. The philosophers of this era (585 BC to 400 BC) worried about reconciling the conceptual mismatches between

a) one and many
b) permanence and change
c) reality and appearance
d) existence and non-existence.

These philosophers had a strong life-centric outlook. They noted that individuals are made from many cells; each species contains many individuals; collectively life on Earth symbiotically controls its own environment; the whole universe contains separate aggregations of matter on different scales. These philosophers puzzled over: What constitutes a meaningful whole ('one')? When is the whole ('one') different from the sum of its parts ('many')? Does the presence of parts signify an imperfection in the whole? This duality also covers Zeno's paradoxes of plurality that, in the absence of more formal mathematics, find the infinite sum of infinitesimals perplexing[2].

A flowing river is in constant motion; the water passing one spot constantly changes yet the river maintains its form. A person ages, changes their size and appearance over time and even exchanges all the cells of the body, yet whilst alive the person maintains continuity of self and purpose. Thus the early philosophers pondered over: What do we mean by the same thing ('permanence') at different times ('change')? How is continuance preserved under pressure of change? Are flux and stasis different aspects of the same thing?

Parmenides developed rational objective arguments that supported the contention that there can be no change or spatial differentiation in reality. He argued that one couldn't describe what is not unless what is exists. Thus what is must always have been in the same unchanging eternal form. His arguments[3] rely on the classical-Greek view of a universe of definite matter occupying a separate space-time. Clearly Parmenides' conclusion is contrary to the subjective appearance of change. This contradiction generated the rational concern that all of experience is unreal! The philosophers queried: Does an objective world ('reality' as we know it) exist at all? How can one decide what is truly objective ('reality') and what is subjective ('appearance').

The final duality from pre-Socratic times, concerns how we define existence rather than what actually exists. To Pythagoras, algebra and numbers (as an eternal system) represented an aspect of divine truth; the expression of this aspect becomes contingently corrupted in the experiential world. In science our basic definitions (rather than empirically derived laws or even scientific principles) were taken to have this same supernatural status. Thus philosophers realized that they could not decide what features of the universe should be given the status of existing until they could decide: What do the terms existence and non-existence mean? What are the essential qualities that denote existence? Is existence an attribute or an independent quality?

The advent of modern science has refined and expanded these historical dualities.

4.2 Modern Dualities

Step 4 involves identifying the most pertinent dualities and considering the mechanism by which the conflicting ideas can be joined. Several such dualities confront common-sense experience. Some of these dualities have been tackled by quantum mechanics and relativity in their various guises. Some dualities are awaiting resolution by the new reality and other dualities may potentially remain for future resolution. Over aeons, our forebears have debated the clash of dualities based upon the following still-potent differentiations:

1. discrete and continuous
2. statistical and definite
3. matter and action
4. local and holistic
5. space and matter
6. qualitative and quantitative
7. mind and matter
8. randomness and meaning.

Note that many of these different dualities may now be seen to be related to one another – ultimately they are all related. In addition, each duality contains lesser dualities. Thus the 'statistical and definite' duality includes the pairing of free will and determinism, it includes the disagreement on the structural basis of reality between a definite classical worldview and a statistical quantum worldview and it includes the partially revealed mathematical correspondences between statistical quantities (eg temperature) and non-statistical quantities (eg complex-imaginary time) invoked in semi-quantized versions of general relativity.

Furthermore, a previous apparently resolved duality might still contain some further mileage to be extracted. For instance, general relativity claims to have already unified space and matter; however, this duality has some current potency because general relativity remains outside the quantum worldview. A good example of a duality that has been previously recycled is provided by the duality of inertial and gravitational mass. Following the triumphs of general relativity, in our list of dualities, we have regarded this old duality as having been subsumed into the space-matter duality.

To accomplish step 4, we are requested to select one duality (or more if they seem to be related) that houses the greatest distinction of concepts. We want to identify the duality that possesses the greatest resistance to being reformulated as a union. Each of the listed dualities retains some conflict. However, current science offers partial resolutions for the first five dualities. The final duality queries how in a truly chaotic universe we can sustain meaning. This is deeply meaningful and an answer is provided in part 2. However, this duality will not be selected; the existing scientific paradigm is almost oblivious to the subterranean dangers lurking in this duality.

In contrast dualities 6 and 7 are very interesting. Yesterday's science (prevenient to eventual reality) is a monarch in the realm of the quantitative and of matter, a monarch who cannot extend her authority into the lands of the mind and the qualitative. These dualities do fit our requirements, they are related, they have greatly resisted unification and (if they are not disabled) they do have the power to be serious obstacles for expanding the scope of reason beyond science.

4.3 Qualitative and Quantitative

Personally experienced reality is predominantly qualitative; in contrast, science only has highly developed tools to describe the observational quantitative realm. Since science defines reality, almost by accident, science has proclaimed the quantitative nature dominant over a subordinate qualitative nature. Thus qualitative concepts within science are regarded as representations of quantitative comparisons. For instance, acids and bases are useful qualitative concepts derived from quantitative comparisons of pH (potential of hydrogen) levels. Unfortunately many qualitative concepts[4] do not fit this style and are thereby outside of the scientific realm.

If we are to reformulate reality in qualitative terms (as requested by step 4 of the map), we need different tools to describe the qualitative nature of reality. Fortunately, since personal experience is predominantly qualitative, our societies already have developed a tool that effectively meets the task's requirements. This tool is everyday language. Language is a sophisticate symbolic system that is used to describe all of qualitative reality including both science and language itself.

Now, our human languages are unnecessarily ugly and ambiguous. The universe will have its own special langue but fortunately we can work with a translation of the langue. Thus we have the capacity to attempt a novel reformulation that conflates the qualitative and quantitative realms.

4.4 Mind and Matter

It is one of the great strengths of science that science provides an objective reality – a reality that we can all agree upon. However, the detail of this objective reality is untrustworthy since (by past experience) even the most beautiful of scientific theories is likely to have a use-by date. Scientists must always be humble; the universe is the ultimate arbitrator whatever the wishes of philosophers, theologians, business managers, reporters, celebrities, public-relations experts, generals or politicians.

Conventional scientific wisdom holds that the physical brain can support activities that lead to mental processes. Hence, it appears to be obvious that the material world is somehow more fundamental than any mentally constructed world. However, this view has not enabled us to comprehend the mind. We have little idea as to what a mind is or why it should exist. As scientists we need to be prepared to overturn our traditional beliefs. Is mind a product of matter or vice versa? Is there 'power' in belief? Can one mind interact with another mind without physically disturbing intervening matter? Accepting the dichotomy of mind and matter as a duality will require us to attempt to merge these twin concepts in a more fruitful way. This merger will require that our customary view of matter be seen to be just one limiting aspect of a richer concept.

Such a reformulation would seem to fit with the qualitative-quantitative duality. Thus we will chose to progress to the next step with these two related dualities. Naturally a qualitative reality may have a connection with thought. Thought is normally associated with a form of language. Note that even when we do not think using our regular vocabulary we still think in symbols ie a form of language.

4.5 Conclusion

Step 4 of the map involves consideration of relevant dualities and the selection of the most divisive ones. This is an important step as the subordinate concept of a duality points to a mechanism from which we shall describe the basis of a new reality.

We have chosen to consider the divide between qualitative and quantitative. We have included with this duality one of the most incomprehensible mysteries for science, namely the mind-body duality. The implications of adopting these dualities are quite profound: the description of the quantitative universe is to be built from a qualitative language; this approach is unlike existing science which describes its concepts mathematically.

Notes

[1] See Shand (1993) pp 3-6 for background on pre-Socratic philosophers and their concerns over dualities.

[2] Note that the most important dualities of each era appear paradoxical. However, not all paradoxes are associated with dualities and not all dualities give rise to paradoxes.

[3] Parmenides argued that there can never have been only nothing in the universe (nothing is meaningless without a comparison to something). Consequently, matter did not spring from nothing – matter has always existed. Hence the fundamental form of matter is unchanging.

[4] For example, concepts such as *meaning* are not amenable to quantitative comparisons.

Chapter 5 : Context

5.1 Introduction

Organisms evolve in an environment that they help to form and mould. Likewise, scientific theories evolve within a shifting worldview to which they contribute. These theories both can respond to changes taking place in the worldview and can challenge the worldview. However, the extent of compatibility of the theory with the acceptance criteria of this worldview and of scientific method determines whether the theory can survive in the market place of ideas or whether it must combat the worldview.

Acceptance is critical for the progress of any theory. Newton was smart enough to propose classical mechanics and not to propose quantum mechanics. Equally we need to understand the forces that are currently shaping and maintaining the worldview. When quantum mechanics was proposed, it was within a classical worldview that was already being challenged by the evidence of the internal structure of the atom and evidence of the interaction of matter with light. These pictures helped to mould a primitive first fledgling form of quantum mechanics that could be grudgingly accepted. Likewise, it is important that our radical theoretical next development be cognizant of and adapted to the current dynamics of the surrounding worldview.

Thus the final preparatory step (before we attempt to assemble a picture of a new foundation for science) involves:

1. identifying those competing, emerging and possibly non-embedded big picture concepts that sit uncomfortably within or along side the current worldview. These concepts provide our direct competition.
2. considering the relevance of consistent trends that have repeatedly appeared in physics. These trends can be used to help justify our shift in perspective.

5.2 Blemishes in our Understanding

We know enough about nature to ask many awkward questions about space and matter, that we are unable to answer unambiguously.

Context

5.2.1 Spatial Dimensions

Particles have more degrees of freedom than those simply expressed by their position and momentum coordinates. These additional degrees of freedom are revealed by the particle's attributed quantum numbers (intrinsic spin, charge, isospin, etc.). This extra freedom has led some to hold the view that there needs to be additional spatial dimensions. However, as direct experience seems only to show three spatial dimensions (plus time), these proponents of extra dimensions assume that the proposed extra six or seven spatial dimensions[1] (numbers that help the mathematics work) are coiled up on a small scale and further entwined in a complex manner. The scale for these coiled dimensions is often imagined to be the Planck length or a little larger. In the quantum philosophy, if we could make measurements on the Planck-length scale, quantum effects should overwhelm the predictable continuum gravitational structure of the space-time continuum. Any scale remotely close to the Planck scale is beyond direct study for the near future.

We cannot know that these extra (Planck-sized) dimensions exist; however, it is assumed that further dimensions are needed to explain particle quantum numbers and to explain symmetry.

5.2.2 What is Matter?

In the current quantitative worldview, we don't quite know what description we ultimately should attribute to the structural elements of matter. Newton established a worldview of point particles interacting at a distance. Heisenberg retained point particles[2] and classical interactions in the early quantum mechanics. In later quantum field theory, we retain conceptual point particles even if (for relativistic purposes) they may only interact locally and even if they are normally smeared throughout space. This is the basis of the highly successful standard model of particle physics.

Unfortunately, although the theory can be made to work, the localized nature of a point particle introduces infinities into many relativistic analyses. Hence, consideration has been given to basing theories on spatially extended objects. Some early theories involved one-dimensional lineal entities known as strings. A later generalization of the theory incorporates p-dimensional entities and this theory is referred to as M-theory[3]. M-theory is held in great regard by many particle physicists because it is hoped the theory will be able to adequately unify (general) relativity and quantum mechanics; because the theory requires the existence of spin-2 massless gravitons; because the theory explains the fermion and boson classes of particles and because the theory can predict Hawking's black hole entropy formula.

However, M-theory is actually a class of theories. It is unable to predict a unique set of coupling constants for the strengths of the forces. To date M-theory hasn't been able to predict the masses of the elementary particles and it requires many more fundamental particles than have been observed (for instance, each known particle is required to have a 'super-partner' with spin that is ½ a unit less the known partner). M-theory is so elaborate that its attractiveness and sensibilities may stem from some but not all of its features. At its core, M-theory has finite-size p-dimensional objects (p-branes) avoiding self-energy infinities, it relates the families of fundamental particles to the excitations of these objects, and hence to the dimensionality of space-time; it requires extra dimensions in the space-time continuum.

In passing, we can note that M-theory has many equally sophisticated competitors. Disappointingly all of these theories are incompletely formulated. These hypotheses seem to be awaiting guidance from experiment to help shape their formalisms. Another substantial theory that has been worked upon by many physicists is loop quantum gravity. Eventually any of these theories could form alliances with our emerging theory of eventual reality. However, one thing is clear; we do not know how to describe matter.

5.3 Competing Explanations

It is worth noting which existing theories are weak. In the cosmological arena, evidence is necessarily sketchy and there appear to be several interlinked theories that may have little genuine experimental support.

For instance some views of reality promote many-world models with our world being determined by anthropic selection. Anthropic selection is a recognition that any world in which we could live has to be special eg having the capacity to evolve 'intelligent' life. Thus if there were many different worlds we probably live in a world belonging to a very special subset of those universes with intelligent beings. Anthropic arguments are valid within the context of a justified model of reality that requires a multitude of redundant universes. However, it is usually unwarranted to apply anthropic selection to an unsupported model so as to generate supporting evidence for the model. Such an activity can only stifle useful research; the anthropic principle cannot be quarantined; everything that we don't know about the universe should not be a target for an anthropic argument!

Some anthropic arguments mark a departure from the way that science has traditionally developed. Science has been nourished by observation but scientific theories are tested by experiment. The attractiveness of an anthropic conclusion should not lead us to rely upon a scientific method based solely upon limited viewer-biased (ie mono-world) observational data, since we would be unable to forbid the use of the principle in even other less-sound areas.

For instance there are separate anthropic arguments that complex evolved life can only exist in a space with 3+1 macroscopic dimensions and in such as space with fundamental constants remarkable close to their currently measured values. We should not object to these particular arguments on the basis that they are circular. However, we should object that these arguments are not falsifiable – unless we can access some consequence of an alternative universe.

In principle some anthropic arguments attempt to make probabilistic predictions about the likelihood of living in a universe resembling the observable universe. These arguments adapt to the scientific framework. Of course if there were an infinite set of universes the chances of living in any one particular universe should be negligible ($1/\infty$) and the chances of living in a similar (identified by coarse-graining in parameter space) universe would either be indeterminate[4] (∞/∞) or negligible (eg $15000/\infty$). Baring these serious technical difficulties the additional scientific concern with probabilistic predictions is that whatever its statistical weighting, a single data point (from our universe) cannot be used to adjudicate upon the value of a probability and hence this statistically unreliable single result does not support or contradict a (non-zero) probability.

Other competition for our new theory may come from the cosmological model of inflation. Cosmological inflation was invented to explain the flatness and homogeneity of space. Cosmological inflation predicts an average mass density close to the critical density and predicts a nearly uniform microwave background and predicts an approximately scale-invariant spectrum of perturbations from homogeneity. On the downside, there is no standard model of inflation, the origin of the perturbations is not well-characterized and there is a counter argument (based upon the limited size of causally connected regions) that the inflation model is incompatible with a non-zero cosmological constant. Thus in the absence of detailed experiments on cosmological scales, all aspects of cosmological inflation are questionable.

5.4 Tidal Currents

Within evolutionary eras, natural selection can impose consistent pressures, eg groups of competing animals might get progressively larger, faster, fiercer or smarter. Likewise, the evolution of science contains discernible trends. When Maxwell unified the previously distinct fields of electricity and magnetism, he would not have suspected that this would be viewed later as the first step in a unification scheme for all fundamental forces (most of which he wouldn't have known existed). Today this trend is often cited to explain the direction of some of modern physics.

However, there is a different evolutionary trend that is less commonly acknowledged but of even deeper significance. This trend has to do with why and how we keep building 'invalid' world pictures. Physics has been described as an intellectual structure built up from broad empirical laws, as a structure explained by theories and as a structure understood by systems of principles. However, there is evidence that this familiar structure of subsuming tiers is founded upon an illusion. It may be that laws are somewhat illusory; nature is less regular, less restrictive, less confined than the imposition of any non-trivial law.

It seems that directly and indirectly, the macro world has provided us with a set of premises as to how to describe reality. However, we keep finding that our notions of reality need amending and replacing. We once thought that light travelled in straight lines. Diffraction shows us that this belief is an illusion. We once thought that energy and momentum were always conserved. Quantum tunnelling shows that this is not the case. Indeed (according to quantum mechanics), in small-enough time intervals and space regions, there is no limit to how large the non-conserved discrepancy in total energy and momentum can become. Macro-world notions can become very entrenched in our thinking; for instance, we have used them to build up laws of reflections and refraction and to build up the theories of relativity. It can prove difficult to disregard such apparent truths when their predictive power has been demonstratively proven useful.

Now, an understanding of how these passé macro-world notions arise is provided by applying Fermat's principle[5] of stationary phase to the less ordered micro-version of reality that lies beneath each of these examples of supplanted truths. The physical explanations are all similar. In a chaotic micro-system where everything seems to be happening, most contributions to any macro-process will nullify other contributions to the macro-process. However, a few contributions that are close to an extreme action value will interfere constructively and provide a practical quantitative calculation scheme involving more regulated concepts. This is how our current laws and associated concepts acquire their credentials. This procedure may be where many of our challengeable notions of reality such as the existence of fundamental particles with fixed quantum numbers, may have their origins.

Clearly the more fundamental description of reality would be that description that is drawn from the more chaotic underworld not drawn from the description embellishing a clinically more regular macro-world. We can glimpse features of this underworld by identifying what is forbidden in our current worldview. Indeed this precept, the use of Fermat's principle to connect different versions of reality, exposes the reason why (in a strict sense) we keep getting our picture of the physical world 'wrong'. This precept explains why

- a ray theory of light was so successful
- a theory of interactions via forces was so successful
- the classical world-view was so successful
- the field of thermodynamics is so useful

when technically each theoretical description is wrong!

Context

There are many things that routinely we take for granted. Conservation laws flow from symmetry. However, conservation laws only apply on the macroscopic level. We habitually say that energy and momentum are always conserved; yet, we allow small-scale cheating and borrowing. Really we should acknowledge the microscopic standpoint that it is more appropriate to believe that energy and momentum are not conserved quantities. It only looks as though these quantities are conserved when we inspect a system on a macroscopic scale. On the macro-level, the stationary-phase contributions dominate and provide our distorted view of the universe. Now, if we don't hold to conservation of energy and momentum why should we hold to any other conservation law such as conservation of spin or baryon number? Clearly we are in need of a more fundamental theory to answer this question as to what quantities are conserved in the stratified descriptions of reality.

The logical conclusion of these trends would be there are not any laws which restrict or describe activities. Ultimately everything is allowed to happen. The corollary is that everything is possible! However, nearly everything normally cancels, so nearly everything is unimportant. Perhaps every imposition or law that we, at anytime, hold to be true is an illusion.

5.5 Conclusion

In this chapter we have considered the environment of ideas influencing the acceptance of our new theory; however, we have not attempted to summarize the status of fundamental physics. Instead, we have highlighted some current issues of importance and highlighted an explanation for a submerged undercurrent.

Thus we identified several weaknesses and serious shortfalls (in our overall understanding) that it would be profitable for the new hypothesis to tackle. The shortfalls stem from a lack of an underpinning foundation firstly for space and subsequently for matter. Space and matter both have intrinsic properties – both physical entities must be derived concepts. Yet since we do not have a more fundamental explanation of reality, we cannot explain space's dimensionality or matter's most fundamental form.

Further, we noted an evolutionary trend away from classical reality and quantitative law towards descriptions based upon greater freedom, upon greater choice and upon greater indefiniteness. The new theory can profit from this trend; indeed, step 2 has already suggested a chaotic basis for reality.

Notes

[1] Although most recent models of space-time (devised by string theorists) assume that there are 10 or 11 dimensions, there are earlier models with higher numbers eg 26 dimensions.

[2] The wave-particle duality allows states to behave as either particle or waves. In the extreme case, when viewed as a localized entity, quantum particles are taken to have no internal structure and to be point like.

[3] Our lack of knowledge of this theory is such that the 'M' in M-theory stands for mystery or occasionally anything else an author wants to propose: magic, make-believe, mumbo-jumbo...

[4] This is an issue with the countability of infinite sequences. The problem demonstrates that one needs a scientific mechanism to generate an infinite multitude of universes and hence be able to uniquely extrapolate from finite sets of universes.

[5] Fermat's principle is explained in section 6.3.

Chapter 6 : New Vision

6.1 Introduction

In the preceding five chapters of this document, we have executed the instructions listed in the map as outlined in step 1 of the map. This activity has generated the revealing constituent parts of our new contending view of reality. A summary of the activities and outcomes of the steps of the map have been listed in table 6.1. For completeness we have also summarized the outcomes of the final steps that follow.

Table 6.1

Summary of the map of paradigm shifts and the map's conclusions about the next evolutionary changes required to the worldview of reality.

#	Title	Action	Result
1	Map.	Follow quantum path	Update map
2	Dark Clouds	New additional concepts	Chaos, complex numbers
3	Master Principle	Universalize additional concept	Unspecialized numbers: complex & quaternion
4	Dualities	Concepts resisting merger	Quality and quantity Mind and matter
5	Environment	Trends, competitors, allies	Less order, hierarchies from Fermat's principle
6	Picture	Assemble around elemental	Qualitative foundation
7	Hypothesis	First Approximation	Multitude of event-pointers
8	Presentation	Document work	This document

The purpose of step 6 of the map is to congeal the conclusions of the previous steps into a consistent qualitative picture of reality. Then the next two steps will convert this picture into an exploitable description.

Let us consider the summary of conclusions displayed in table 6.1. In step 4, we deduced that the underlying conceptual basis for the new view of reality will be qualitative; nature is to be fundamentally qualitative– but with a subservient capacity to express quantitative features. Step 2 and step 5 suggested that the foundation of everything rested upon a total lack of constraint. Step 5 suggested Fermat's principle was a useful tool for sculpting a hierarchy of conceptual structures each with seemingly greater classicality. The first use of the principle would be in building an elementary quantitative world from the qualitative foundation. In step 3, we exposed the master principle that assigns numbers that are moderately unspecialized to the description of a derived quantitative reality. In step 2, we identified mathematical-complex numbers as one of those unspecialized numbers. It was suggested in step 3, that quaternions were possibly the only other such unspecialized number required.

This set of results from our journey provides an explicit picture of the important concepts of the emerging hypothesis. We are even told where to begin in creating the new hypothesis of reality. Our strategy is to ask: What should we accept as ultimately real? We have already rejected the classical and quantum concepts of a reality founded upon particulate matter. Instead, we have adopted some elemental qualitative entity as the fundamental basis of reality. Hence, we grow the theory from a qualitative seed. This assembled picture of reality is quite simple: fundamental reality is describable by a qualitative chaotic schema upon which a simple quantitative reality in unspecialized numbers will eventuate by application of Fermat's principle. Turning this picture into a fully-fledged hypothesis (to be known as eventual reality) is left to the next parts of this document. However, a start is made in the following section in which we firm up the character of this picture.

6.2 Quale

Our map has not yet given us a description of the essential elemental quality that is to form the basis for all reality. This objective is both the last part of this step and the initial part of the next step. A separate chapter is not provided for step 7; thus we will start this step's explanation here. We continue to use our meta-principles as guides with the added proviso that we are not to overburden the new theory with any unnecessary further concepts that have not already been selected in the previous steps of the map. Further we are instructed to be prepared to keep temporarily some old concepts in order to help the transition between the old and new worldviews of reality. Primarily we are encouraged to be simple (theory of fresh eyes/Ockham's razor). We are encouraged to form the new theory around the concept of the qualitative.

Now, classical mechanics formalized the corpuscular hypothesis; quantum mechanics has retained this hypothesis as a model component. We will mirror this corpuscular schema and assume that qualitative reality can be assembled from a multitude of elemental entities. These entities are describable by their essential qualities: they have no size; they have no shape; they have no mass; they have no quantum numbers; they have no position in time or space; they have no quantitative relation with any entities. There seems to be little that we can attribute to these entities. However, ultimate qualities must derive their character by being relative[1] to or dependent upon other qualities. Hence, pointing (ie referencing other entities) seems to be the primary property and probably the only property that we can allow. Note that these entities do not point to some place nor do these entities point to something physical; they point to other similar entities that have the same inclination simply to point. The more-developed attributed qualia are to be formed from this relative arrangement of pointing.

The new hypothesis developed here is to be known as the theory of eventual reality and this hypothesis will be described in some detail in the next parts of this document. The linguistic term *eventual* can mean ultimate but in archaic use, the term meant possible (cf. eventuality). Both of these two meanings are appropriate although the relevance of the first meaning was not apparent when the theory was tentatively christened. The theory subsequently imposed its name on the ultimate entities possessing only qualia; these entities are called *events*. These events are timeless, spaceless, qualitative elements; their only function is to point to other events. These events should not be confused with occurrences or circumstances in space and time.

Now, the terminology *monad* is frequently used to describe the unique fundamental form from which everything is build. Thus events can be described[2] as monads possessing the characteristic qualitative functionality of referencing other events.

6.3 First Principle

Fermat's principle is fundamental to eventual reality. Indeed it will become apparent that Fermat's principle is the fabled first principle of the universe from which all the studies of knowledge and the recognizable universe arise. The importance of the principle is such that even though the principle will be familiar to most readers I shall take this opportunity to discuss in some detail what is meant by the principle in this work.

Depending upon the nature of the system to which the principle is applied, we shall observe that different aspects of the principle are revealed.

6.3.1 Classical Stationary Weight

Classical stationariness notes the importance of the weight (ie statistical measure) of incoherent configurative contributions that are to be summed into a resultant. The most likely configuration is stationary. Image that these contributions are describable by one-dimensional Dirac delta functions ie the ultimate peaked contribution. Now, when we add (ie integrate) these contributions the resultant need not possess any turning values.

$$W_R(q) = \int_{-\infty}^{+\infty} dr\, W(r)\delta(q-r) \qquad (6.1)$$

The resultant function, $W_R(q)$, has a turning value only where the weight function, $W(q)$, has a turning value. In this example this conclusion is a truism as the two functions are numerically equal. However, within the resolution spread of any contribution function, the conclusion is quite general applicable to any sum of contributions. This classical stationary result is a way of expressing the relation that when the weight function is maximal, the resultant is stationary.

Furthermore the weighting function can steal the physical relevance from the underlying contribution. For instance, a gas in a one-dimensional box has a statistical-mechanical number density

$$n_{SM}(q) = \sum_i \delta(q - x_i) \qquad (6.2)$$

but the averaged thermodynamic number density, $n_{th}(x)$, looks like the weighting function

$$n_{th}(x) = \int_0^L \frac{dq}{L} n_{SM}(q) = \frac{N}{L}. \qquad (6.3)$$

We will consider one more gedanken example of the classical coherent weight. Suppose we set up a camera to randomly take photographs of ourselves on a swing. To a reasonable approximation, the probability of the camera snapping the swing around angle θ in the exposure time Δt is simply

$$P(\theta)\Delta\theta = \Delta t / \mathrm{T}_p \text{ where } \Delta\theta(t) = \theta(t+\Delta t) - \theta(t) \qquad (6.4)$$

and where T_p is the swing's period of oscillation. Thus the probability density, $P(\theta)$, is given by[3]

$$P(\theta) = \frac{1/\mathrm{T}_p}{\Delta\theta / \Delta t}. \qquad (6.5)$$

Thus the most commonly observed angles are when you, the swinger, are physically stationary at the turning points of the motion. These turning points denote the angular extremes of the swing.

6.3.2 Classical Stationary Phase

Fermat's principle was proposed as an explanation as to why a ray of light took the path of shortest travel time between any two points on its trajectory. The lessons about the impact of a stationary weight function apply but, in this example, we are presented with contributions that interfere both constructively and destructively. The stationariness of the weight of a contribution is now to be interpreted as a depiction of symmetry. For simplicity we shall assume that a scalar-wave theory may be used.

Away from the stationary condition, the phases vary wildly; these non-stationary contributions add to a seemingly incoherent background noise. Often the multitude of contributions will render the background relatively negligible. Around the stationary condition, the contributions are almost in phase locally and thus the contributions add coherently. Hence the resultant appears to be formed from dominant stationary contributions that add coherently ie at locations where a set of contributions that are in phase.

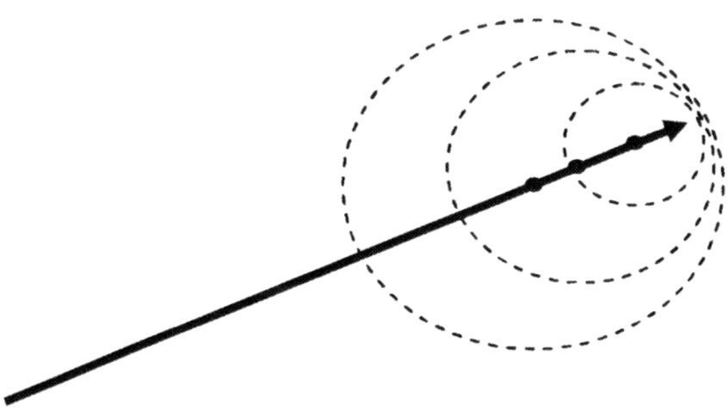

Figure 6.1

Phases of Huygens wavelet contributions from earlier in trajectory coherently add to perpetuate the trajectory.

Let us consider the original question of the path taken by a ray of light. The Huygens' wavelets originating along the prior path of a ray constructively add in the forward direction perpetuating the form of a ray. In a homogeneous medium, symmetry precludes the stationary condition[4] for a ray being anything other than a straight line.

Clearly homogeneity is important and a ray description is only an approximation. Recall that a destruction of symmetry by a reflective diffraction grating shows that somehow the ray of light is present elsewhere along the grating. Reflection of a ray by a mirror satisfies the law of reflection: angle of incidence equal to angle of reflection. A ray theory predicts that the mirror is only required at one point in space; however, wave theories correctly tell us otherwise.

Likewise, we use the same symmetry arguments to predict that the Huygens' wavelets from a spherical wavefront will propagate a spherical wavefront in the forward direction. These two examples underscore the basis for the duality of wave and particle[5] concepts.

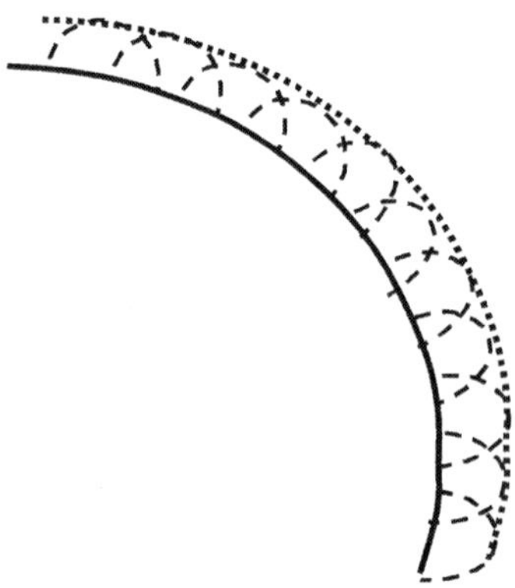

Figure 6.2

Phases of Huygens wavelet contributions from the wavefront coherently add to perpetuate the wavefront.

Thus very often the prior symmetry takes charge of the expressed macro-resultant propagating form – whether that form takes the shape of a wavefront, a trajectory or an intermediary.

6.3.3 Eventual Stationariness

Fermat's principle can be applied at an even deeper level of reality. At all levels, Fermat's principle is a statement of both of logic and mathematics. It is independent of the physical system to which it is applied. The most general expression of this principle is given without an externally identifiable weight function and without an externally imposed symmetry. These important ancillary features can be identified as home grown from within the prescription of the stationary condition.

Thus the eventual stationary condition is to be written simply in the form

$$\delta S = 0 \qquad (6.6)$$

which depicts a universal stationariness of an unspecified assembly of contributions. Logically this stationary condition is not a statement that the summation, S, is a universal constant; rather it is a statement that S is described by a stable resultant structure not a fleeting form. The stable structure identifies the coherent character of the dominating set of contributions. Furthermore the dominance of the stable structure is cemented by the sheer abundance of contributions. Inevitably this equation will have a solution because we cannot face a reality that is totally incoherent. Indeed total practical incoherence, as an ideal state, may not be possible. As the classical Greek Philosopher, Parmenides, would have argued: incoherence is a lack of coherence; incoherence is not something in its own right. Thus we cannot have incoherence unless we also have some coherence from which the incoherence can be differentiated. Hence we should not be surprised that there will be a solution when we apply Fermat's principle to the eventual representation of reality.

In this equation we now treat S as stationary for infinitesimal variations of each coherently reified variable that S depends upon. By design the summed solution will depict a coherent resultant displaying symmetries from which properties can be derived. These symmetries betray the nature of the freedom of the unseen contributions[6]. If there are a large number of coherent contributions, then the incoherent sum will be relatively unimportant. This unimportance may be taken to the limit of an infinite number of contributions.

6.4 Conclusion

Without mentioning the map we could have presented a range of supportive arguments for the development of this expressed picture of the theory. However, the task of forming a new paradigm was made simple and straightforward by breaking the task up into simple component tasks. A large leap is much more difficult than a series of small steps.

Now, the map describes an inductive process; following a long scientific tradition we have attempted to unravel nature's secrets by inference from observation. Thus only a small degree of faith is needed to accept that the suggested tasks will provide a hypothesis that is useful to investigate. However, a greater degree of faith is required to maintain that the hypothesis will be useful explanatory tool, let alone maintain that the hypothesis is correct.

However, even if the hypothesis proves unsatisfactory, the methodology of the map may have further benefits elsewhere in science. Furthermore, in the light of lessons from a partial failure of this approach, the method could be repeated to generate an improved hypothesis. Clearly we are now in a position to take this derived picture of a new reality and develop the picture into a working hypothesis. Thus part 2 will describe a solution to steps 7 and 8 of the map. Finally part 3 will describe another complementary version of eventual reality.

Chapter 6

Notes

[1] The fundamental building blocks are to have no unjustified properties. Hence they acquire their qualities through their inter-associations.

[2] A list of linguistic terms is provided in the preface.

[3] An average over a large number of swings of similar amplitude is implicit in this equation.

[4] A homogeneous medium has no inherent directional bias. Thus the bias built into the boundary conditions is perpetuated.

[5] In this picture, a particle may be viewed as the propagating front point of the ray.

[6] One may picture the sets of individual contributions in many ways. However, it is the set of conditions (or lack of conditions) imposed upon the ensemble of individual contributions that is important in fixing the resultant.

Epilogue to Part 1

Hopefully you have enjoyed this gentle downhill stroll across the deeper parklands of physics, a stroll that has illuminated many details about the nature of reality. The climb back to conventional reality will be more strenuous; however, on our downhill journey we have glimpsed many of the landmarks that will guide our route back to the dominion of experience. Furthermore we have also discovered useful climbing gear in the form of Fermat's theorem.

However, before we begin this construction of a new scientific foundation we will mull over some selected philosophical matters that put in perspective what we have accomplished in part 1 and also set the scene for our account of the new formalism in part 2. There are certain questions that have slippery answers: What is Physics? What is human? What is science? What is real? In every era the answers to these questions are set by a different convention. These questions invite consideration of the likelihood of an ultimate reality. An ultimate reality, if it exists, should be devoid of convention; we expect this ultimate reality to assign invariant truths. Indeed, if there is an ultimate reality and the methodology described in part 1 works, then our procedure should be able to uncover this ultimate reality after a finite number of iterations.

We expect this ultimate reality to be able to convincingly answer compelling questions like: Why does the universe exist? What is the essential characteristic of existing? What are the fundamental entities of reality? What is the nature of the fundamental entities of reality? How do the fundamental entities build our contingent observable universe?

However, we will never be sure of the existence of such an ultimate reality until our scientific theories make contact with this reality. Thus in part 1, we have built a picture of our new theory by following the scientific tradition and we have relied upon experience to guide a construction of the generated viewpoint. In this way we have kept indirect contact with empirical reality. We have steered away from purely rational arguments uninformed by experience.

Not surprisingly in the battleground of ideas there are many competing arguments over the appearance of fundamental reality. It is likely that there is a smidgen of truth in all of them. Nevertheless collectively the arguments serve only to confuse, as the following incomplete list will help to illustrate.

Some (like Pythagoras) would argue that the fundamental representation of reality should be pure mathematics. After all, mathematics is supposedly a field of logical universal truths. Furthermore if we compare different scientific disciplines, we may crudely use the mathematical sophistication and percentage content of mathematics as a measurement rule to identify the degree of fundamentality of the scientific discipline. Naturally physics is the most fundamental science. Congratulatory trends are always appealing but we should not surrender theoretical physics to the mathematicians. Furthermore it is not apparent that the same correlation applies across the different fields of physics. For instance both classical mechanics and solid-state physics are intrinsically highly mathematical yet philosophically less fundamental than elementary particle physics.

Some would argue that this fundamental representation should be based upon perfection. Thus ultimate reality should be immutable and invariant. Any change or variation in true reality would reflect an imperfection. After all if only what is can exist, then what is not (eg an absence of something in time or space) cannot exist. Then, in actuality, space should be full ie space should be a plenum without prospect for material movement.

Some would argue that this fundamental representation is an indefinite repetitive collection of all possible realities. Then, nothing that we could know or come to understand about observable nature could represent anything but biases placed upon the detectable 'monoverse' by our awareness only within this monoverse.

Some would argue that this fundamental representation of reality is constructed from one unique self-consistent form from which everything else is a specialized instance. After all, the number one represents simplicity and may be almost self-evident. In contrast a number, like forty-two, would seem to demand a much more substantive justification.

Some would argue that this fundamental representation of reality represents a balance of opposing fluxes. A balanced universe is an impartial universe; a universe of fluxes implies that motion is more fundamental than substance; thus this conclusion suggests that ultimate things are self-formed from the essence of the dynamics.

Some would argue that this fundamental representation of reality is rigidly constrained by restrictions that fix the unique laws and properties of the observable universe. Yet others would argue that there is nothing to impose such constraints and hence there are not any constraints.

This incomplete list is illustrative of the inventiveness of the mind. The extent of the veracity of such arguments will become clear in part 2.

Part 2:
An Exposition on the Theoretical Hypothesis of Eventual Reality

Prologue to Part 2

Part 2 reports upon the development of a comprehensive hypothesis known as eventual reality. Part 2 contains the crux of the theory and has been prepared so that it may be read independently of part 1. Thus some readers may initially skip part 1 which describes how the hypothesis was conceived but does not describe the formalism of the hypothesis.

Here the hypothesis of eventual reality is presented as being derived from a small number of tenets. These tenets are used to construct a multistorey edifice of conceptual reality. From the vantage point of the storey on which individual consciousnesses reside, these foundation tenets can be shown to be self-evident. Thus eventual reality is shown to provide a philosophically consistent description of reality.

The character of eventual reality displays a radical departure from the traditional scientific worldview. Thus to prepare the reader for the arguments in the following chapters, I will provide an outline of the general appearance of the theory. The surmise of eventual reality is that fundamentally we inhabit a universe of countless options, a cornucopia of potentiality. These options form the basis for every aspect of fundamental reality; they deliver a foundation description of the universe existing independently of any quantitative space and time. Indeed this foundation does not display any of the familiar concepts such as energy, matter, consciousness or free will; there is only endless possibility. The options are rendered discussable by formulating a model representation of the universe in terms of an infinite set of pointers (the pointers are referred to as events). The pointers reference other pointers; they are entirely qualitative and individually they carry no sense of distance or direction.

We shall explain that this network of pointers is identified with the fundamental universe; however, the network does not correspond to any form of space-time. This foundation description occupies the ground-level storey in the theoretical edifice that attempts to explain all. We should recognize that the pointers are countable and (via Fermat's principle) the need for symmetry imposes an overarching structure on the collective outcome of the total assembly of options. Thus we are able to build the first-elevated storey of reality with an explanation based upon the dominance of effective conspiratorial coherence amongst the options. This first-elevated storey reveals a generalized quantitative coordinate space with hierarchical sub-network connections (referred to as an aethereal field) that are representable as expressions in a universal langue.

The aether sits below our recognized reality. Interpreting the aether in terms of a rich structure of localized informational entities builds our viewpoint of matter in a hybrid space, which is obtained by transformation of the foundation coordinate space. The reality of the aether and the reality of matter are quite different. Thus, unlike hidden variables that predetermine outcomes, the aether is seen to provide the statistics for an ill-informed classical-style quantitative reality. In this second-elevated storey, we observe how matter can combine with the aethereal language to fix choices in a time-propagating manner. In all, we discuss six storeys in the conceptual edifice of reality each storey with a unique picture of the universe. As is to be expected, the distinction between neighbouring storeys becomes less discernable at the higher levels.

The theory of eventual reality is designed to embrace more than quantitative science. On the scientific side, the theory explains why there are regular analytical laws of nature. We show that this foundation leads to known physics. For instance, it is demonstrated how quantum-mechanical principles appear naturally within this theory. We give a physical picture to many stray scientific concepts that were previously without an *ab initio* basis. The theory is used to successfully derive the symmetries of our space-time continuum and the symmetries and quantum numbers of elementary particles.

However, the greatest power of eventual reality is in its capacity to explain much that was beyond science. We show that this picture of reality is self-evident. We answer many of the great mysteries of philosophy. We explain the nature of mind and we obliterate the mind-body problem.

Before we commence the detailed exposition of the theory of eventual reality, I should prepare the reader for the style of several important arguments. Most readers will be familiar with the powerful apagogical technique known as *reductio ad absurdum*. In such arguments, we prove a premise false by initially asserting that the premise is true and then deducing a logical contradiction. The logic of eventual reality will oppose the use of such arguments. In contrast, we shall use the rarely favoured technique of *by inspection* . This technique confirms the compatibility of a statement with an asserted premise that is within the purview of the statement, by presuming the premise to be true. In this manner, the technique shows that the asserted premise is an acceptable solution of the statement. For example in mathematics, it may be possible to solve for a solution of a quadratic equation in x by assuming that a solution is $x = 2$ and then confirming by substitution that this solution satisfies the quadratic equation. Quite generally the *by inspection* method produces circular arguments in which we assume that which we wish to confirm; there is no indication of uniqueness of the solution; there is no indication that unacceptability (for other reasons) of the solution will imply falsehood of the governing statement. However, almost by default, when investigating the most fundamental aspects of reality we are forced to accept this unconventional logical technique.

Part 2

Chapter 7 : Introduction to Eventual Reality

In this work we suggest an alternative to the conventional worldview of physics. Part 2 of the document presents an outcome of a practical and systematic analysis of the development of scientific worldviews; this analysis of worldview development was given in part 1. The new theory has many appealing aspects but the theory's formulation is still in its infancy. However, the theory does demonstrate how it is possible to construct a complete[1] theory of reality that is consistent with the observations of modern science. Clearly some or all of the arguments and interpretations presented here may prove to be naïve or even totally misconceived. However, every aspect of science should always to be open to challenge by rival hypotheses. There is much further work needed to develop the details of the theory.

This new hypothesis is referred to as the theory of eventual reality. Eventual reality proposes a quite different way of comprehending the measurable aspects of the universe including both space and matter along with their properties and dynamics and a new way of describing the immeasurable effable quandaries of the universe such as consciousness and the subjective. The provision of an opportunity for scientific assessment of the new theory dictates that this introduction to the hypothesis will focus on the implications of eventual reality for those physical 'things' that are measurable. However, the perspective of consciousness cannot and will not be ignored.

The theory of eventual reality is an example of a reformulation of science that aims to displace the materialist structural foundations that are held by most current theories. Part 2 will show that eventual reality is capable of being the progenitor of science's most successful theory, quantum mechanics. Furthermore, eventual reality makes some clear testable quantitative predictions such as definitively describing the measurable structure of space and time. This deduced structure for space-time differs from that structure proposed by other current theories.

Ultimately eventual reality proposes to supersede all the current fundamental theoretical representations of nature that form a collection of mechanics, geometric theories and models. Mechanics (such as quantum mechanics) deal with interactions and the motion of matter in a given spatial metric. Geometric theories (such as general relativity) deal with the topology of space and gravity, treating space, matter and events within space as purely relative. In addition, a multitude of models (such as the standard model of particle physics or quintessence in cosmology) provides descriptions of posited elements of reality that are to obey the principles outlined by a mechanics or geometric theory or both. The current bizarre collection of accepted theories lacks a unifying conceptual formalism in which they can be embedded. This set of eminently useful accepted theories also contains many fixed numerical values (often considered as constants) that must be supplied from outside of the theories. For instance, these fixed values may include the masses and coupling constants of some elementary particles, the scale of space and time and an array of factors (covering cosmic inflation, free-space energy density) influencing the expansion rate of the universe. The perceived paucity of the explanatory capabilities of the standard worldview have led some researchers to seriously consider various multi-world models with anthropic selection principles[2] that serve to place much of the totality of the universe beyond direct observation. Such approaches cannot be ignored simply because they are an anathema to many scientists. Nevertheless such principles unnecessarily and prematurely undermine science's playful and arrogant belief in its eventual capabilities. Certainly models of reality ought to be appropriately established before we accept the conclusions of anthropic arguments applied to these models.

Eventual reality proposes an explanatory hierarchical structure for current physical theories and their theoretical progenitors and progeny. Indeed, it even proposes a scheme (outlined in part 1) by which new fundamental representations of science (and its successors) might be progressed. Furthermore, in total, eventual reality uses fewer basic concepts than the suit of theories that make up the mainstream scientific worldview. Not surprisingly a lesser number of basic concepts imposes a greater number of restrictions upon how reality can be constructed and hence this stronger limitation makes eventual reality a 'rich' (ie strongly deductive/predictive) theory. Thus eventual reality can provide explanations for the existence of the known laws of physics and their form. For instance, eventual reality explains the principles and rules of quantum mechanics. The theory requires that (on particular levels in the hierarchy of theories) the universe display specific symmetries. For instance these symmetries describe a space-time continuum at odds with the 10 or 11 dimensional space-time proposed by some string models and their successors[3].

These definite symmetries provide one avenue for testing the propositions of eventual reality. In this part of this document, it will be shown how the symmetries of eventual reality provide a physical interpretation of elementary-particle properties such as electrical charge, hadron-colour charge, elementary-particle families and rest mass.

Thus the hypothesis of eventual reality is more than speculation; the hypothesis draws definite testable conclusions. The hypothesis purports to explain every aspect of reality from just a small number of self-evident tenets. These tenets will be presented and then the elaborate multi-storey edifice, which is the framework of a conceptual understanding of nature, will be developed. It is only when we reach the third-elevated storey where our consciousness resides that we can appreciate why the foundation tenets are self-justifying.

It is important to understand the motivation and focus of this report. Some readers may be disappointed with the limited development of the hypothesis in those area most of interest to the reader. The presented work has a narrow aim; the aim is to generate a new theoretical foundation for the universe with an associated comprehensive picture of reality. As such, the analyses necessarily build the scaffolding for the series of storeys that make up the hierarchical edifice of reality. The focus of the report is upon this construction of the framework, rather than upon illuminating fine details of the structural building blocks of all of reality. For instance, a simple form of quantum mechanics was primarily derived to reveal how a particular storey of reality was constructed; the purpose of the derivation was not to reveal the principles of quantum mechanics.

Notes

[1] It is proposed that eventual reality will explain every aspect of the universe eg the existence of science, the nature of the theory of eventual reality, the necessity for individual intelligences, etc.

[2] Rees (1997) pp243-269 presents a good account of the range of anthropic reasoning. There are many hypotheses that employ anthropic arguments: a fascinating evolutionary example was given by Smolin (1997).

[3] Greene (1999) gives a popular account of the ideas and requirements for a universe of p-branes.

Part 2

Chapter 8 : The Motivations for Change

8.1 Why Displace Quantum Mechanics?

Quantum mechanics has inbuilt limitations placed upon what it can describe and to what it is applicable. By acknowledging some limitations, quantum mechanics' claimed scope for its worldview is more restricted than that scope originally claimed by classical mechanics. For instance, quantum mechanics only allows the attribution of the results of quantitative measurements on otherwise isolated subsystems of the universe. As we shall now discuss, this restriction has implications related to the principle of determinism.

Firstly let us recall the infected wound inflicted by determinism upon classical mechanics. The Marquis de Laplace noted that if classical mechanics were precise and applicable to all matter then there should be no free will[1]. One might suspect that quantum mechanics could overcome this deficiency through the indeterminacy of the predicted result of a measurement. However, quantum indeterminacy is too limited to achieve this goal of permitting any independent subsystems to have sentient choice. The reasoning is straightforward; sets of choices and actions are not free if they are constrained even by a probability function.

Unfortunately this undesirable attribute is not the most serious deficiency with quantum mechanics – quantum mechanics is not even internally consistent. Essentially quantum mechanics is constructed from two mutually contradictory parts: the first part contains time-symmetric deterministic principles describing the propagation of a quantum state; the second part contains associated regulations describing a time-asymmetric and indeterminate measurement procedure. The quantum principles (providing the equation of motion) are applicable to an isolated subsystem of the universe. The rules of measurement are applied when this isolated subsystem interacts with a measurement apparatus. Incompatibilities arise when we realize that the isolated subsystem and the measurement apparatus form a larger new isolated subsystem. When the complete experimental system is viewed as an isolated subsystem coupled to a measurement apparatus, quantum mechanics requires that the result is statistically determinate and that some details of the pre-measurement state are irretrievably discarded by the measurement procedure. However, if the entire experimental system were viewed as one large isolated subsystem, quantum mechanics would require that determinism holds and consequently that we could reverse time to recover the precise starting state. Thus these two parts of quantum mechanics are mutually incompatible. The two parts of quantum mechanics cannot both be *precisely* correct. Intriguingly there would be no incompatibility were we to apply quantum mechanics to the entire universe. The universe is perfectly isolated and there can be no external measurement equipment. Hence, only the principles of quantum mechanics could be applied and (if the universe were to obey quantum mechanics) quantum mechanics would require that the universe must be fully deterministic.

Furthermore quantum mechanics suggests, in many other ways, that quantum mechanics isn't a fundamental theory. For instance, quantum mechanics is only a component of an analytical mosaic structure of interconnecting theories of science. Quantum mechanics describes the dynamics of a system; however, additional models of reality describing matter and space are required before the theory can be applied. Quantum mechanics doesn't select these models (eg the standard model of elementary particle physics) nor has quantum mechanics been able to choose our universe, the universe's laws or fundamental constants from the wide choices that seem to be available.

Finally I note that since the founding of the Copenhagen interpretation[2], some physicists[3] have held the view that quantum mechanics must be an incomplete theory. Firstly no satisfactory interpretation has been given for why quantum mechanics does yield only the statistics of outcomes. Secondly a statistical description is an unusually undecided way to describe what is supposed to be all that can ever be known about the investigated system. Indeed, historical suggestions of incompleteness have been linked to the 'relativistically challenged' nature of distant measurements[4] on an entangled quantum state of isolated particles (ie the Einstein-Podolsky-Rosen paradox). Probabilities and statistical correlation functions are presented classically as describing statistical details of an ensemble of systems; ie these functions of the ensemble do not describe the details of any one particular state. We can have an ensemble when there is further unregistered information. If such classical statistical precepts are to be replaced, quantum mechanics should either provide the replacement rationale or accept the label of incompleteness.

Nevertheless we must acknowledge that quantum mechanics is supremely successful at predicting outcomes. There is no suggestion of any limitation in quantum mechanic's predictive capability, from any experiment test. However, the arguments, given in this section, highlight the conclusion that quantum mechanics is bereft of a reliable philosophical basis and a sound logical basis. These dilemmas have been unresolved for 80 years. Numerous modifications to quantum mechanics have been investigated. However, after so many years, it is reasonable to propose that further tinkering by making small adaptations will not succeed and that major changes to our total conceptual foundations might be due.

8.2 Why Rewrite the Paradigm?

Science and philosophy definitively parted ways with Newton's formulation of, what amounted to, a clockwork universe. Indeed this comprehensive formulation marked the birth of modern science. Unfortunately Newtonian-style determinism is incompatible with our intuitive belief in the autonomous free will of individuals. Initially it was hoped that either there were two immutable forms of matter: one organic and the other inorganic or there were two immutable forms of existence: the form of matter and the form of an immaterial soul. In either case, one form would obey classical determinism and the other form would be subject to different freer criteria. However, nowadays we accept that classical mechanics is only an approximation. Hence, the accuracy of classical mechanics is limited; therefore the *rigorous* deterministic capabilities of classical mechanics cannot be utilized to determine the philosophical basis of nature. Thus, for example, classical mechanics (because it is an approximation to nature) does not preclude the possibility of free will.

This chasm between an acceptable science (agreeing with experiment) and an acceptable philosophy (agreeing with logic and necessary experience) remains for all the major theoretical developments of the twentieth century. Hence, like the modern view of classical mechanics, it is the better option to regard these twentieth-century theories also as approximations rather than as precise predictors of the natural world. Now, there are two mainstream scientific paradigms of physics: one paradigm is centred on the theory of quantum mechanics and the other paradigm is founded upon the theory of general relativity. We have explained the philosophical conundrums of the quantum worldview in the previous section. General relativity has a purer philosophy; it is founded upon the assumption of a general invariance of physical law. However, general relativity retains much from classical mechanics including Newton's curse of determinism. General relativity has other failings but a detailed discussion of these failings will be left for part 3 (sections 15.2 and 16.1.2). There have been numerous unsuccessful attempts to marry quantum mechanics and general relativity. Current attempts (if successful) to marry these two theories are likely to import the philosophical problems of quantum mechanics into general relativity. Thus it seems that neither of these central theories (nor even the legendary combined theory, the theory of everything, formed by simply merging the rival theories) will have a fully satisfactory philosophy. Consequently we are led to conclude that *both* theories (as currently understood) are only approximate and we are unlikely to complete the merger of the theories until we can comprehend a lower common stratum of reality.

Equally any new theory with a contradictory philosophy should also be regarded as approximate. However, it is not necessary to regard all future theories as approximations. Our existing paradigm needs replacing because the paradigm contains contradictions between the nature of its prediction, its requisite philosophy and common experience. An aim of scientific study should be to heal this incompatibility by devising a theory that does not have to be wrong ie devising a theory without any contradictions. It transpires that eventual reality is such a theory.

The mainstream scientific paradigms hold the view (material monism) that we live in a material-based world. Actually we inhabit the world of our own minds. Furthermore the nature of the mind has steadfastly defied attempts at a scientific explanation. The mind-body problem can be considered as a series of questions without clear answers: Is a mind a consequence of matter? Is matter a consequence of mind? Are mind and matter different aspects of the same thing? Are mind and matter derived consequences of something else? Are minds contained within space-time? What is the nature of consciousness? What is the nature of free will? Is all matter conscious? How does the mind both receive inputs from matter and control matter? Our existing scientific paradigm does not provide a framework that can answer these questions. Thus the existing paradigm is deficient and the paradigm needs updating or replacing.

Inevitably modern science brilliantly describes much in the material world. Yet, surprisingly modern science does not explain satisfactorily many of the fundamentals of this material world. We cannot explain the requirements for the dimensionality of the universe, the nature of fundamental particles, the description of dark matter and dark energy. We cannot yet explain without ambiguity how the universe began or how/whether it will end. We cannot explain clearly the nature of time, the imbalance between left-handed and right-handed matter, the quantity of matter, the size of the universe, the fundamental constants of nature, etc. Indeed, we don't even know which physical 'constants' are universal invariants and which physical constants are dependent conditionals.

Hence, it could be argued that modern science regularly explains the mundane but modern science is severely limited in its scope of applicability. Indeed, it is likely that all fundamental concepts are beyond the current scientific paradigm.

If this conventional worldview founded upon pictures of matter and space is unsatisfactory, we should be prepared to entertain alternative pictures that could be much simpler and more elegant.

The theory of eventual reality is not simply another mechanics describing the dynamics of matter. It is a theory that seeks to define and explain reality and thus eventual reality seeks to substantially rewrite our entire worldview. This theory exposes the foundation of reality. The theory reveals a universe built upon an uncontrolled expression of a single absolute ideal universal form.

8.3 The Origins of Eventual Reality

Of course one of the strongest reasons for attempting to create a new understanding of the universe comes from wanting to test whether the conclusions of part 1 are useful. Now, there are two commonsense but strikingly different routes to presenting a formulation of the theory of eventual reality: one route would be from observations of nature (based upon evidence and empiricism) and the other route would be from logical arguments about necessity (based upon reason and rationalism). As a fundamental theory, eventual reality can be presented using either of these approaches (starting from the top and working down or starting at the bottom and working up). The universe is as we observe it to be but the universe is as it must be. Both routes should agree and, within the author's knowledge, they have only been shown to agree within the framework of eventual reality.

For instance, we shall show (in section 13.2) that the theory of eventual reality can be given an *a priori* justification that is philosophically sound. Philosophers from before the modern era were able to recognize many of the characteristic features of eventual reality's philosophy because the science of those earlier times was not sophisticated enough to blind them to appropriate lines of argument. However, because their scientific knowledge was limited, these early philosophers were unable to develop their philosophical ideas into a comprehensive paradigm without introducing unknown mystical elements. Today, the problem is reversed. We know enough about science to develop these arguments adequately. However, our prejudices are so coloured by the apparent lessons of rigorous invincible science that it is extremely difficult to weigh the relevant philosophical judgements and difficult to see the *wood for the trees.* It is for this reason that the philosophical foundation of eventual reality will not be discussed until chapter 13, ie until after the paradigm has been unveiled in this document.

Actually, in an indirect fashion, eventual reality was developed from observation of nature and observations of science. This procedure was described in part 1 of this work. Part 2 is intended to be somewhat self-contained and readable independently of part 1. Hence a short passing description will be given as to how the hypothesis was developed.

The initial development of eventual reality arose from general considerations of the way in which progress is made by science. This study of science's development is imbued with a series of expected beliefs about science. These beliefs were drawn from the myths of science. These simple beliefs were applied to our knowledge of the natural world, the existing scientific paradigm and our stories of scientific exploration. This application generated an explicit set of explicit instructions for forming a new scientific paradigm. This set of instructions is referred to as a map; the individual instructions are referred to as steps of the map. This map provides a slightly different view of the values of eventual reality from the view presented in part 2. The map's values are closer to an identified subset of those values held by the current mix of scientific theories than the values that arise in the formalism. As such, the values espoused by the steps of the map are to be regarded as derivatives of the inherently fundamental values, as presented here, of the hypothesis of eventual reality.

Thus part 2 will present the fruits of applying the derived picture of reality, thereby, demonstrating the utility of the methodology presented in part 1. Hence, to present a digestible first introduction to the theory, this report turns the conclusions of part 1 around and presents the theory as though it were derived from a formal structure incorporating a set of premises and propositions, which are known as the tenets of eventual reality.

In the picture provided by eventual reality, the elemental unit of the universe is termed an event. It could equally have been called a monad. This event is a pointer (to other pointers) but the pointer should not to be confused with a point or location in space-time. There is no space and time for the individual events, only references to other events. These pointer functions are qualitative; pointers have neither an intrinsic sense of direction nor a sense of separation. These pointers embody choice as to where they point; and qualities can be attributed to events, based upon the environments to which the events point. Just as matter is generally regarded as being comprised of discrete particles (or perhaps structures of higher dimension), the universe (not to be confused with the space-time continuum) is presumed to be describable by an infinite ensemble of events. It is proposed that space and matter are derived composite quantities of this universe. Such features represent a coherence exhibited by the universe much like the macroscopic thermodynamics variables that display the combined effect of many microscopic molecular influences or like the macroscopic electromagnetic variables that reflect microscopic-system responses or like the ray-optic trajectories that are the dominant macroscopic residue of very many microscopic wavelets. These changes of descriptive picture that lead to the derived concepts such as space and matter can be understood as examples of a generalization of Pierre de Fermat's principle of stationariness. 'Macroscopic' classical mechanics has this same derivative relationship to the principles of 'microscopic' quantum mechanics (in a path integral formulation). It will be shown in this work that 'macroscopic' quantum mechanics bears the same derivative relationship to 'microscopic' eventual reality.

A hypothesis in which the universe is built from such 'events' is believed to provide the simplest model that is consistent with the inherent underlying inductive values of eventual reality as revealed by the evolutionary map. These values hold that that the entire universe is a valid concept, that reality consistently displays simple universal elegant explainable principles, that there is an inherent duality between the qualitative and the quantitative.

This picture will be further explained and developed into a practical theory in the next chapter.

Notes

[1] The debate about the restrictions of determinism goes back much further in time at least to the ancient-Greek proponents of atoms, Leucippus and Democritus.

[2] See section 1.2.2 for an introduction to the Copenhagen interpretation of quantum mechanics.

[3] It is well known that Einstein was most uncomfortable with quantum mechanics. He felt that, by introducing probabilities, quantum mechanics was turning its back on physical reality genuinely existing in space-time.

[4] When a quantum measurement occurs, our knowledge of the system changes and the state is required to change instantaneously. However, simultaneity is a relativistic variant.

Part 2

Chapter 9 : A Description of Eventual Reality

Please be aware that we are time dwellers. Thus the vocabulary of our language is ill-equipped to describe a timeless basis for reality. However, rather than coin new words, in explaining the picture provided by the theory of eventual reality, I have chosen to misappropriate existing words that conventionally have connotations of temporality, to convey a state of latent decidedness.

9.1 The Tenets of Eventual Reality

The presumptive starting point for the theory of eventual reality can be expressed in terms of a small set of (somewhat vaguely described) tenets. The full relevance and meaning of these four foundation tenets will become clearer as they are discussed and applied. Allied forms of most of these tenets are reasonably familiar in science. Ultimately these tenets will be interpreted in section 13.2 as the *jus naturale* of the universe. These tenets define the foundation storey of reality – fundamental reality.

Premise 1: Every possible option is expressed.

This simple statement of complete freedom disguises a lot of hidden detail. The statement is not meant to be applied directly to our familiar everyday reality. The statement represents the idea that the foundation of all of reality is derived from the superposition of an unlimited number of choices possessed by something so much freer than everyday reality. This premise is more far-reaching than the statements: *'Everything happens'* and *'No process is to be forbidden'*. However, when everything does happen, then this freedom is such that another opposite cancelling contribution almost always accompanies every contribution to an action, observation or choice. Nevertheless we shall observe that the immensity of the uncertain chaos can accumulate so as to exhibit a discernible pattern. Thus we can be left with a simplified coherent more macroscopic[1] picture of reality that is understood as the dominating resultant behaviour of this milieu.

Let me remark that we shall study this coherent picture by applying Fermat's principle of stationariness[2]. Interestingly Fermat argued that nature was 'economical' always choosing the quickest or the shortest route (the original theorem is known as principle of least time). Pierre-Louis Moreau de Maupertuis developed the idea of an economical dictate of nature into the more general principle of stationary action and, in a later guise, the principle morphed into the variational principle of classical mechanics. I shall simply refer to this principle as Fermat's principle of stationariness or occasionally Fermat's principle of stationary phase. Note that in modern usage, we treat nature as profligate, with those seemingly economical (or more correctly stationary) processes representing the remnant larger-scale coherence of the resultant summation of copious smaller-scale contributions.

Premise 2: The foundation of reality is qualitative.

This requirement is meant exclusively – there is to be no room for the quantitative in our foundation description. Now, there is a startling divide between the quantitative and qualitative aspects of direct experience. Conventional science (with a paradigm steeped in the primacy of the material world and the rule of mathematics) has not been able to extend its realm of applicability beyond the physical quantitative world. Philosophers have long defined at least two other figurative worlds[3] these other worlds are the world of personal experiences and the world of abstract knowledge and argument (including numbers and mathematics). The quantitative picture of reality seems to have a paucity of potentiality to describe the dual quantitative and qualitative nature of our experienced reality. Indeed, there seems to be no reason why any mental experience should exist in a rational material world.

Actually a scientific theory can be part quantitative (predictive) and part qualitative (explanatory). Thus it is not surprising that current (quantitatively focussed) science does not explain why the universe is so understandable ie why science should exist. Rationally a quantitative reality cannot be completely elementary. Everything (other than possibly some elements of abstract mathematics) that is quantitative is a composite union of a quantity and a quality. Thus part of quantitative reality will always depend upon qualitative values (ie qualities). Although, from a materialist perspective it is not immediately obvious, the contrary dependency need not be the case.

Thus the hypothesis of eventual reality starts with the surmise that the building blocks of reality are not the measurable things of quantum mechanics[4] but that the building blocks of reality are non-numeric entities. From this qualitative beginning, we shall deduce that quantitative things, which are describable by mathematics and amenable to objective investigation, can be said to exist. Furthermore, *composite* qualitative entities are based upon and ultimately are elements of a general language, the langue of the universe. Intriguingly such qualitative entities are capable of spawning the material, the mental, the measurable and the functional.

This qualitative foundation gives the theory of eventual reality an internal self-consistency, enabling the theory to describe both physical nature and natural philosophy. This premise is the starting point for an explanation of the metaphysical. Thus eventual reality is a theory of natural philosophy. This document, however, will focus mainly on the objectively testable physical implications of eventual reality.

Chapter 9

Proposition[5] 1: The principles of natural philosophy are hierarchical in nature and built using Fermat's principle upon an increasing semblance of order.

It is noted that successful scientific theories fit within a relational hierarchy. This hierarchy is roughly based upon scale[6] but actually governed by a generalized version of Pierre de Fermat's principle of stationariness, which is instrumental in successively changing the basis of our descriptions of reality. In amidst a lack of constraint (premise 1), we shall identify coherent behaviour and we shall attribute properties and permanence to this coherence. The coherent behaviours betray the created symmetry of the descriptors of this rung in the hierarchy.

Electromagnetism and statistical mechanics both contain microscopic and macroscopic descriptions. In the same vain, classical mechanics can be thought of as a macroscopic version of quantum mechanics. The apparent irreconcilable conflicts within or between theories can be viewed as being a consequence of theories being on different storeys of this hierarchy (referred to as the edifice of reality). In this hierarchy, quantum mechanics is a macroscopic form of the coordinate-space[7] representation of eventual reality. To help develop the theory, a gravity-free continuum version of eventual reality will be described in part 2. Simplified descriptions of the curved space form of eventual reality are left for part 3.

Proposition 2: All of reality (including those parts that we recognize as describing mind and body) is ultimately built from qualitative elements that solely reference and relate as pointers to other similar qualitative elements.

These elementary qualitative units of the universe are termed events. They are the fundamental monadic forms. No event is intrinsically more special than any other event. However, events are distinguishable according to their local network connections. Note that these elements are *not* discrete 'atoms' of coordinate space. However, events can be taken to be nodal pointers in a more general universal network that does *not* constitute a (finite-dimensional) space. In this statement we recognize that an infinitely dimensioned space is an oxymoron[8].

The other general tenets of eventual reality have sufficient loftiness and vagueness to be quite consistent with mainstream systems of scientific belief. However, this last proposition has the specificity of a model (describing the peculiarity of the system's specification). Now, it is common for a hypothesis to construct its fundamental building blocks from discrete elemental units. In this regard, proposition 2 is similar to many familiar hypotheses. However, while discrete units of space-time are part of some debated hypotheses and conjectures, the event is asserted to be the indivisible unit of something else, namely the universe. The event has few describable features; the event is entirely relational and hence qualitative. Thus the event is viewed as a pointer to other events. Furthermore, a pointer event is subtly different from a node in a provided network. The pointer events create and control the links that form a network; the events do not belong to a geometrically assignable place in a network – these pointers are not like vectors, the network is not located in an external space. All events are referential, the network that the events form, and that we will make quantitative by counting (sets of linked events) is inherently relative. All the subsequently assigned qualifications of an event are consequences of the immediate network referenced by the pointer values attached to events.

Thus proposition 2 is a key distinctive feature of the theory of eventual reality and, as we shall see, this proposition offers clear possibilities for direct testing of the theory. Note that within the hierarchy of descriptions proposed by proposition 1, proposition 2 has limited direct applicability! Proposition 2 describes the unit of reality on one theoretical level only. Nevertheless the proposition's implications are widespread in the derivative descriptive higher storeys of the edifice of reality.

These four tenets describe the core belief structure of the theory of eventual reality. All omissions from the list are as important as these described inclusions. Note that this list of tenets included no external presumptions about many commonly accepted principles such as determinism, causality, analyticity, materialism, continuity, uniformity, spatial number forms, fundamental symmetry, etc. Thus by current standards, eventual reality has a very lean foundation and the traditional scientific worldview is substantially more presumptuous.

Furthermore, after a simple form of the theory has been explained, rational explanations of all four of these tenets will be given in chapter 13.

9.2 The Foundations of Space

The menagerie of conventional physical theories catalogues many descriptions of traditional space-time. Usually such spaces are taken to be coordinate-system manifolds that provide a quantitative framework for describing where things are and how they are arranged relative to other things. In this section, we shall consider the eventual-reality view of the ancestral coordinate space from which all of quantitative science is to be derived. Let me offer a caution: this ancestral coordinate space is not the same as the hybrid space of our physical awareness; hybrid space will be described in chapter 10.

We will build a sequence of conceptual representations to be referred to as storeys of the edifice of reality (proposition 1). Within the theory of eventual reality, coordinate space (situated on the first-elevated storey) is to be a coherent feature of a fundamental universe that is constructed from an indefinite arrangement of interlinked events. This view of the foundation of reality is formed by applying the model presented as proposition 2 and by exercising premise 1. Recall that events are defined as relational pointers that point to other events. Each pointer can point to any number of other pointers.

However, the pointing is not to be without relevance (ie not pointless), these pointer assignments are to be the basis for connections between the events. This linked network of events is known as the apeiron.

Thus we are about to unveil the secret to the conundrum: how to conjure something (eg a partially quantitative universe) out of *thin air*[9]. Once this *sleight of hand* procedure is appreciated we should find that visualizing the transition to each successive higher storey of reality will become easier and easier. However, this first stage, which erects a framework to stand upon our founding tenets, is potentially the stage that is most challenged by our prior accepted beliefs. In order to reorient the thinking of the reader we shall discuss the principles that we are about to use, with examples taken from familiar reality. This discussion should help to prepare the reader for the fuller arguments that are to be used to expose the necessary quantitative properties of the foundation space. Hence in the next sections we shall explain how and why a foundation coordinate space is formed. Importantly we shall discuss the relationships between coherence, symmetry, stationariness, invariance, groups and algebra. Subsequently the arguments will be revisited and then specific conclusions about the nature of coordinate space will be derived in section 9.3.

9.2.1 Invariance and Symmetry

On each storey the recognizable features of the universe are a mix of certain arrangements that are contingent upon circumstances, and of non-contingent necessities. For instance, in our perceived 3-dimensional world, to describe a contingent shape such as an isolated sphere or rod, generally it is easiest to choose coordinates that match the shape's symmetry even though this local symmetry does not fit the (contingent) arrangement of matter in the rest of the universe. However, we could choose any coordinate system and we could place the origin of this coordinate system at any location with any particular orientation. These coordinate-system choices are arbitrary, they may affect the description of coordinate locations but the choices do not influence a set of invariants belonging to the relativities of the system being studied eg in a Euclidean 3-world the length of the rod does not depend upon where the origin of coordinates is placed. In this way we note that this available freedom of the coordinate frame describes a characteristic symmetry (of the non-contingent aspects of the associated reality) that becomes enshrined in the (non-contingent) hierarchical laws and principles of the science of this storey of reality. Contingent information, such as when is the next full Moon, is not fixed by universal (non-contingent) scientific law although this occurrence may be calculated from such scientific law with suitable contingent input data. Note the implied interconnections between invariance, symmetry and freedom; these words are all being used to describe different aspects of the same thing. A specification of the invariance under a non-contingent symmetry transformation is a specification of the permitted representations of the symmetry and in turn is a specification of the available freedom of the quantitative coordinate representation.

Next, we will consider the algebraic representation of this non-contingent symmetry.

9.2.2 Representational Groups and Symmetry

The character of a displayed spatial symmetry can be represented by a mathematical group. A set of spatial coordinates form a group under a binary operator $*$ provided that 1) whenever any two of the variables in the equation

$$\eta^{(i)} * \eta^{(j)} = \eta^{(k)} \tag{9.1}$$

are members of the group, the equation will have a unique solution with the remaining variable also being a member of the group and 2) the action of the binary operator is associative

$$\eta^{(i)} * \eta^{(j)} * \eta^{(k)} = (\eta^{(i)} * \eta^{(j)}) * \eta^{(k)} = \eta^{(i)} * (\eta^{(j)} * \eta^{(k)}). \tag{9.2}$$

Let us consider the generic implications of the first 'completeness' equation (9.1). If $i = k$ then the group must possess a right-identity element. Similarly if $j = k$ then the group must possess a left-identity element. Then, if we make element k an identity element, we observe that every element has a corresponding inverse. The associative character of the operator provides a compatibility that delivers a consistent symmetry (associated with operator $*$) and ensures that, although the order in which we place the coordinates can be important, the order in which the operations are performed is not important.

To see why we need the associative condition (9.2), we shall consider a transformation, T, which we can apply to a variable, η, in the form $\mathrm{T} * \eta$. Suppose that we wish to be able to factor the transformation into two terms such that the transformation is equivalent to first applying T_1 and then applying T_2, ie

$$\mathrm{T} * \eta = T_2 * (\mathrm{T}_1 * \eta). \tag{9.3}$$

Let us apply the transformation to the identity element. We observe that

$$\mathrm{T} \equiv \mathrm{T}_2 * \mathrm{T}_1 . \tag{9.4}$$

Now, we shall apply the transformation to a general variable η

$$\mathrm{T} * \eta = (\mathrm{T}_2 * \mathrm{T}_1) * \eta = T_2 * (\mathrm{T}_1 * \eta). \tag{9.5}$$

This illustrates the 'consistency' requirement which with the completeness requirement provides us with an algebraically representable symmetry[10]. We may also note that without the associative requirement there would be n-1 different n^{th} order terms.

Thus these two requirements form the generic template for a mathematical group. Furthermore the particulars defining which group is being used, are defined by the specifics of the algebraic rule governing the binary operator ie the operand's relational table (eg see tables 3.1 and 3.2).

Both the imposed symmetry of space and the symmetry of coordinate's relational table describe the parameterized freedom available to the coordinate structure. Now, the symmetry operation will act upon and transform all coordinates but the symmetry operation maintains the algebra of the binary operator as an invariant ie if we apply a spatial symmetry operation (eg coordinate system translation) to a coordinate, then the result is still a coordinate. Thus the character of a space can be defined either by its set of simple geometrical symmetries or by a corresponding set of binary algebraic operators.

9.2.3 Coherence and Symmetry

The instruction list defining the linkages within the apeiron is of infinite length (premise 1). Let us subdivide the list into sections. Most potential arrangements of pointer assignments within these partially complete lists generate linked sub-networks that make little sense as an organized coordinate space of any topology and any finite dimensionality. Thus it is possible to imagine the universe not as one network but as the superimposition of many possible interlinked sub-networks (the overwhelming majority of which would not seem to make sense as a space). This is a useful way to consider the problem since we do not need to consider the senseless sub-networks that serve to counteract each other.

An analogy may help to exemplify how the working of an ensemble of sub-networks can be viewed. Imagine a set of Huygens wavelets initiated in a collection of weakly linked three-dimensional chaotically linked sub-networks. In almost all randomly structured sub-networks, these wavelets would rapidly self-cancel with wavelets from other sub-networks. Survival of a sustaining disturbance is limited to certain sub-networks that are accompanied by more than their fair share of collaborative reinforcing sub-networks. This cooperative situation reflects the resulting maxima in the parameterized function generating the statistical weight of the sub-networks.

Likewise, in the universe's interlinked chaotic network, the superposition of the sub-realities of any influence, existing across all the disorganized sub-networks, is likely to be nullified by some other sub-network influence. However, a very restricted subset of sub-networks will provide a dominating structure that can exhibit an observable coherent behaviour. I shall refer to this style of argument as rational topology. These dominating sub-networked structures are those that closely admit an appropriate (as yet undefined) symmetry set. Those sub-networks with approximately this symmetry will add coherently. If the ensemble of sub-networks is described in terms of randomly assigned phases, the coherence of the restricted subset of sub-networks can be viewed as an example of domination by stationary-phase sub-networks. These dominating sub-networked structures will define a recognizable coordinate space which when viewed macroscopically resembles a continuum space.

Understanding the way in which a quantitative reality springs into existence is very important. For readers who find the above argument a little too intangible, we shall redescribe the process in a more explicit but rather over-simplistic manner. We are familiar with the way in which some chemists have traditionally used balls and rods to display complex molecules. We shall construct a similar universe in a potentially infinitely dimensioned Euclidean space. All the rods (representing pointers) will be the same length, no angular data is provided and rods are permitted to swing around to point in any direction. We will build the network according to a linkage prescription that we are given. As we build part of this network, it is to be expected that the number of dimensions in which the shape protrudes will grow as the constructed system becomes larger. However, this need not be the case. The list of instructions is of infinite length, it does not have a supplied starting position, and the list can even be partially reordered. We have started at an indeterminate spot in the instruction list. Fortunately an infinite number of other people have also been given the same list and started at different places. Almost everyone produces different seemingly useless tangles of sub-networks, however, a sprinkling of people will find their section of the infinite sequence makes sense and they generate low-dimensional structures. If we wanted to describe these structures quantitatively, we could do so using the angles that the rods made to one another. In the useless uncompacted constructions, these angles form an almost random distribution. Indeed for the most part these angles are independent of one another. Suppose we represent the directions of the rods by direction cosines. The direction cosines will be correlated only in those sub-networks in which most dimensional components of the direction cosines are close to zero ie when, for practical purposes, we have a low-dimensional space. When most of a sub-network's set of direction-cosine components are negligible, then the larger (ie appreciable) direction-cosine components are found within the volume of parameter space that approximates to the symmetric requirement of a low-dimensional structure. In these sets of symmetric sub-networks, the values of the angles are critical and correlated. Then, invoking Fermat's principle, any low-dimensional spaces that are stationary will dominate.

9.2.4 Numbers and Symmetry

Now, let us observe how we would coherently assemble a quantitative space from the underlying universe of boundless possibilities. Inherently all events have equal (and limited) capabilities. Events are simply pointers to other events. A quantitative space will be describable using coordinates. *Ab initio* there is no special event or special spatial location around which we must centralize a universal coordinate system; likewise there is no special link to preferentially identify any unique direction in coordinate space. Thus we should be able to use the same lexicon of mathematical tools to describe our general structure wherever we locate the origin of coordinates or however the chosen coordinates are orientated. This observation allows us to uncover the acceptable characteristic symmetry of coordinate space. Furthermore, these transformed descriptions should preserve reality whenever coordinate-system rotations and translations are suitably combined and then appropriately reversed. Mathematical groups express the logical rules of such sustainable symmetries. The symmetry requirements that are dictated by rotation and translation operations, define those groups that a foundation space will be asked to legitimately express. Once we have identified a successful solution space, we will rely upon the *by inspection* method to justify these imposed presumptions.

Step 7 of the map asks that we do not overburden our endeavours nor take the new theory too far from the scientific community's comfort zone which is identified with prior conventional wisdom. Hence for mathematical simplicity, we confine considerations in part 2 of this work to those of a flat gravity-free space. Incidentally this flat-space approximation will be replaced by an alternative curved-space approximation in part 3 of this work. Now, the expressed choice of an event's pointer function distinguishes one event from another event. In addition, the pointer function provides a means to logically associate sets of events and thereby provides a means to make ordered sets of events countable. Thus ordinal numbers can be used to provide a basis for applying an overall quantitative scale to be applied to the dimensions identified by the sustainable group symmetries of a coherently formed space. On renormalization, this scale will introduce separation lengths in space; some of the separation lengths may be rational but some of the lengths will be irrational numbers (eg recall Pythagoras' theorem). Ultimately we will also depend upon the *by inspection* method to justify the use of such mathematics and logic in our arguments.

9.2.5 Stationariness and Symmetry

Finally let us reiterate what a stationary space means. The concept of a stationary space is quite unusual. In one sense, the resultant space is stationary with respect to a virtual index labelling sub-networks; each sub-network is assigned an even-weight (ie statistically uniform) distribution. However, there is no observable or definable remnant of this index: the index is not decipherable in the resulting coherent space. Thus practical stationariness is only manifest in the observable symmetry peculiarities of this space. Since space is quantitative (built from counting events) space is describable by mathematics. Hence the space will exhibit symmetries as described by mathematical groups: a partial or broken symmetry cannot be stationary. Contributions must show 'phase' coherence when subjected to a fluctuation in any recognizable variable. This requirement is why I am generally referring to Fermat's principle as the principle of stationariness – rather than nominating a fluctuation against which the universe is to be stationary.

Now, we will be imposing two independent symmetries that will instil translation and rotation upon our space. Hence we expect the quantitative descriptor of this space (eg the coordinate) to be the product of the elements of two mathematical groups of different style. Let us discuss these symmetries in familiar terms. Note that the translation and rotation operators do not share characteristic 'eigen'-groups since translations and rotations do not commute – one group has to be given notational superiority over the other in defining our coordinate system. Now, each group is assigned an independent binary operator. Thus the coordinate is given an argument part (indicating magnitude) and a designator part (indicating direction). By choice of convention (in the mathematics), inferiority will given to the argument group and the argument group's binary operator is called addition, $+$. This convention will yield a Cartesian form for representing the coordinate. Hence the binary operator for the designator elements is given primacy and this binary operation is called multiplication, $\times$. The two operators compete when we apply them to a general coordinate. Arguments and designators do not mix but both will be engaged in a multiplication operation. Hence general multiplication of two coordinates will deliver a new coordinate that may have changed both direction and magnitude. However, the addition operator is applied along designator-identified component lines in space and addition can only changes argument values. Therefore it is convenient to define orthogonal component directions along which we may add components. Furthermore a change in magnitude under multiplication requires that the addition group accommodate multiplication as a subservient add-in operator along the line. Multiplication and addition are distinct, yet addition is the only defining operator for the addition group (ie along the line of mathematical-real numbers). Thus, along the line, we should not be surprised that (scale) multiplication comes to be associated with repeated addition. Nevertheless our coordinate number system attributes primacy to multiplication not addition! By these associations, we come to link stationariness with group impositions upon the multiplication and addition operators.

9.3 Fundamental Coordinate Space

Let us now exploit the principles that we have just discussed. For any spatial symmetry to be exhibited there must be at least one dimension in a resultant coordinate space. For rotations to be expressed there must be at least two dimensions. Firstly we shall consider some symmetry implications for the resultant space in one fixed direction. One-dimensional translation introduces the concept of the line in coordinate space. Recall that there is a one-to-one mapping between the mathematic-real numbers[11] and a continuum line. This isomorphic mapping holds a measurement scale (linked to the frequency of events) and defines binary mathematical operations (multiplication and addition) along this line. These operations are accompanied by reverse operations (subtraction and division) that are equivalent to addition and multiplication with a negated and an inverted number respectively. Such symmetries are necessary to show that a sub-network is stationary. Next, we shall consider how these binary operations (+, ×) are to be generalized from simply acting upon a line to acting within some higher-dimensional coordinate space. Of course if there were no higher dimensions in our coordinate space, then we would have already uncovered the algebra of an uninteresting coordinate space based upon the mathematical-real numbers. This uninteresting one-dimensional space does not admit rotations.

Now, in a Cartesian-coordinate representation, the generalized addition operator is isomorphic to a general spatial translation or displacement operator. In the dominant sub-networks that form coordinate space, rational topology requires that such macroscopic translations will form an Abelian group. The generalized multiplication operator will be treated as the 'product' of a change of scale magnitude and a rotation. If new lines are drawn radially from the origin to a point, a change of scale magnitude can be equated with multiplication on the new line. Thus, from the properties of the mathematical-real numbers, changes of scale magnitude are seen to form an Abelian group. Furthermore rational topology requires that macroscopic rotations within a plane will be Abelian but rotations in intersecting planes will be non-commuting. Hence, generalized multiplication performed in coordinate space, may form either an Abelian or a non-Abelian group (depending upon the dimensionality of space). Finally we may use geometric equivalences to confirm the distributive law. If a line of finite length is cut in two pieces, the sum of the pieces adds to the full length. Thus, if a rectangular area is cut into two rectangular areas, the sum of the areas of the smaller rectangles should equal the area of the original rectangle. This result is a geometric restatement of the distributive law of multiplication with respect to addition. These mathematical relations describe the symmetries of the quantitative aspects of the dominating spatial sub-networks.

Thus we conclude that the dominating sub-networks form structures that correspond to the algebraic requirements of a mathematical division ring. Only the expressed connectivity of these dominating sub-networks (ie those that closely exhibit the required symmetries) can take the inbuilt freedom of the apeiron and constructively conspire to give an overall impression of a genuine geometrical network. This explanation describes the mechanism that introduces the concepts of numbers, of sets, of symmetry and of group theory into the physics of the universe. The specification for symmetries is not axiomatic or arbitrary; the specification was derived naturally by applying the requirements of Fermat's principle.

Let us review this important result. It is the stationary nature of coherent space that is responsible for this expressed symmetry. A stationary sub-network has special symmetries that set the sub-network apart from the multitude of nondescript contenders. For instance, sub-networks exhibiting symmetries without a special dimensionality, cannot be stationary sub-networks.

Together mathematical-real numbers (arguments) and designators (directions) define a coordinate representing a location in the coherently formed space. It is convenient to make this distinction between argument and designator; however, these two parts of a coordinate always accompany one another. Thus (affine[12]) requirements on translation and on scaling determine the mathematical-real number axioms. Rotational symmetries dictate the product-rule axioms for spatial Cartesian unit directions. The Cartesian number form arose by adopting the convention that multiplication would be applicable to both argument and designator parts of the coordinate.

Now, vectors are commonly used to represent the coordinates of spaces. Our symmetry requirements rule out such vector spaces. We can understand this result in two ways. Firstly the mathematical-real-number line reveals the mathematical operations that the spatial coordinates can obey. Generalizing multiplication in stationary networks with associated requirements for inverses provides the acceptable symmetry conditions for a coherent space thereby excluding vector spaces. Alternatively we note that there is no additional symmetry imposition (eg on dimensionality) that can appoint a special vector space. Thus no vector space is stationary and hence we do not form vector spaces out of the coherence of the apeiron.

Many of the group properties that we identified came from the properties of lines which can fan out from one point through any other spatial point. The line gained its scale from the counting of events. Thus the line is a line of number. These numbers (along the line) are accompanied by opposites, inverses, identities, etc. The line is not a vector (number composite); the number line has no sense of direction (other than pointing along the line). Inevitably the final symmetry (mathematical division ring) describes a pure number – a result anticipated by Pythagoras.

Note that the event is a discrete element of the apeiron that describes the foundation state of the universe. Thus the totality of the universe is not to be identified with coordinate space or even identified with things and processes within this coordinate space. Coordinate space is revealed by the symmetric dominating sub-networks of an infinitely more irregular universe of possibility. Only events contributing to the dominating sub-networks may resemble 'atoms' of coordinate space. Furthermore, the multiplicity of sub-networks does not involve a set of independent parallel universes. There is only one fully interactive universe. Every sub-network has equal validity; however, only those (few) sub-networks exhibiting regular space-like symmetries are relevant for most analyses.

The requirement that the symmetries of dominating sub-network are those of a mathematical division ring, unambiguously fixes the structure of coordinate space. Only two finite groups[13], those groups associated with mathematical-complex numbers and quaternion numbers[14] form a division ring over the mathematical-real number field. Both of these symmetries that are commonly designated as SU(2) or Sp(1) (for mathematical-complex numbers) and Q_8 or H (for quaternions), are allowed and there are no further restrictions imposed upon space. Thus by premise 1, both symmetries are necessarily exhibited in resultant dominating spatial sub-networks. Furthermore, *only* these symmetries are exhibited in the dominating spatial sub-networks that form coordinate space. However, there are further special requirements on the mathematical manner in which this dual-symmetry impost is accomplished. These considerations[15] will be discussed shortly.

In deducing this symmetry specification for coordinate space, we seem to have implicitly assumed that a coordinate space should show translational and rotational symmetries. However, having derived this consistent form, we can reverse the argument. In the chaotically networked universe, a symmetric form of this derived type will exist as a coherent space. Then we discover that this form displays rotational and translational symmetries. Thus a coordinate space should exist with these symmetries. Here our pre-knowledge of useful representations is obscuring a subtlety. As the hypothesis of eventual reality is developed, the reader will begin to appreciate that every possible representation and interpretation is equally valid but not equally useful. This fact is not social relativism; the different interpretations are never in disagreement. Importantly we are most interested in the useful representations that can inform and describe. Our (biased) pre-knowledge of what might be useful is valuable in this regard. Thus we should not be concerned about all subjectivity that may enter as we form a hierarchy of storeys of reality (proposition 1).

To help the reader follow the pursuing arguments, I shall identify what these mathematical-complex quaternion coordinate dimensions represent. We might not be surprised if the coordinate dimensions (of this complex quaternion space) were identifiable with energy (ε), momentum ($\underline{p}$), time (t) and distance ($\underline{x}$). Many topics of physics introduce these quantities as basic associated variables. Quantum mechanics pairs position and momentum as complementary variables. Likewise, the Hamiltonian approach to classical mechanics uses a canonical Fourier complementary between position and momentum. Statistical mechanics introduces position and momentum as the coordinates of phase-space. However, there is a contrary viewpoint. Recall that for Newton, mass was the measure of the quantity of material substance within a space. Similarly energy and momentum hold a measure of substance whereas the variables such as time and position do not hold this measure. As Lagrange realized, momentum is not a natural choice for a coordinate variable. However, dividing momentum by an invariant measure of quantity such as rest mass, removes this measure of quantity whilst retaining all the physically noted associated correspondences. In special relativity, this quantity is known as the contravariant velocity. Both[16] quantities are possible conjugates to position. Thus in eventual reality, depending on the units system used, the natural complement of position may be either momentum or this velocity. Following the physical reasoning we shall work mostly with a velocity as the designated position-conjugate variable.

For convenience, we shall renormalize the relativistic-contravariant velocity using the speed of light. Thus we equate the (quaternion scalar and quaternion vector) components of the conjugate-velocity vector to the components of the momentum four-vector divided by the product of the rest mass and speed of light, $m_0 c$. Depending upon the emphasis, sometimes I shall use the term conjugate velocity but at other times, I shall use the terms specific energy and specific momentum to mean the same quantities.

This naming of the coordinate components has been given to enhance the physical clarity of the discussion. No *a priori* properties of, or relationships between, the variables are to be presumed. It is necessary to independently demonstrate all mathematical correspondences such as those between velocity, position and time.

Thus with suitable normalizations (not initially shown), a general coordinate would look somewhat like a mathematical-complex number with quaternion components (displayed here as a bracketed scalar and vector pair).

$$\hat{r}\left(\varepsilon, \underline{p}\right)/\, m_0 c + \hat{i}\left(t, \underline{x}\right) \tag{9.6}$$

In this equation, $\hat{r}$ and $\hat{i}$ designate the real and imaginary parts of the mathematical-complex coordinate. The shorthand notation, in which the identity element $\hat{r}$ of the mathematical-complex group is omitted, is apt to misdirect and this notation will not be used in this representation of the coordinate. Now, writing the conjugate velocity as

$$(w_0, \underline{w}) = (\varepsilon / m_0 c^2, \underline{p} / m_0 c) \tag{9.7}$$

this dimensionless coordinate takes the form

$$\begin{aligned} \underline{\underline{\chi}} = & \left(\hat{r}\, w_0 + \hat{i}\; Ht\right)\hat{\gamma}_0 + \left(\hat{r}\, w_1 + \hat{i}\; Hx_1 / c\right)\hat{\gamma}_1 \\ & + \left(\hat{r}\, w_2 + \hat{i}\; Hx_2 / c\right)\hat{\gamma}_2 + \left(\hat{r}\, w_3 + \hat{i}\; Hx_3 / c\right)\hat{\gamma}_3 \end{aligned} \tag{9.8}$$

where H is a scale factor that compares the relative sizes of the complex-real and complex-imaginary dimensions, c is the speed of light and the $\hat{\gamma}_i$ are quaternion-designator[17] labels for the dimensions of physical space.

$$\begin{aligned} &\hat{\gamma}_1\hat{\gamma}_2 = -\hat{\gamma}_2\hat{\gamma}_1 = \hat{\gamma}_3 \text{ etc} \\ &\hat{\gamma}_1^2 = \hat{\gamma}_2^2 = \hat{\gamma}_3^2 = -\hat{\gamma}_0^2 = -\hat{\gamma}_0 \\ &\hat{\gamma}_0\hat{\gamma}_i = \hat{\gamma}_i\hat{\gamma}_0 \qquad i = 0..3 \end{aligned} \tag{9.9}$$

Essentially equation (9.8) provides the definition of the (position, time and conjugate four-velocity) variables that label the dimensions of this coordinate space. Note that for a free particle with three-velocity, $\underline{v}$, special relativity gives the conjugate velocity as the contravariant unit-length 4-vector

$$(w_0, \underline{w}) = (1, \underline{v}/c)/\sqrt{1-(v/c)^2}\,. \tag{9.10}$$

Explicitly the separate representation of these velocity variables, allows the quaternion-vector components of conjugate velocity to be used to quantify the pace of a change in location and likewise the quaternion-scalar component of conjugate velocity to be used to quantify the pace of a change in time (ie represents the flow-rate of time).

Now, individually mathematical-complex and quaternion numbers form division rings. Yet, mathematical-complex quaternion numbers do not form a single division ring. Thus mathematicians have sought to invent higher dimensionality generalizations (eg octonions, sedenions) with weaker symmetry properties. The inadequacies of directly combining mathematical-complex numbers and quaternion numbers can be readily pictured. In such ill-considered complex-quaternion arithmetic, the square root of minus one $(-1\hat{r}\hat{\gamma}_0)$ could be a mathematical-imaginary scalar or it could be a mathematical-real quaternion vector. This ambiguity provides an unwanted confused mixing of the mathematical-complex and quaternion-vector representations. Fortunately there is a way to avoid this uncertainty.

Now, in a geometric representation, the mathematical sign of a variable signifies a directional sense for a physical displacement. Physically we would not expect that the directional sense of a vector should be equated to the directional sense of a mathematical-complex component. Hence, we could label all mathematical signs with their structural relevance. For this labelling to work in practice, we are required to be able to factor the coordinate space into terms belonging to the separate division rings each with their individual sign operators. We shall adopt the notation that an absent positive, +, sign is implied and thus positive signs can be suppressed. A negative, –, sign will be either grouped with its designator or provided with a subscript to make clear along which direction the sign operates. The outcome for a coordinate is that only five of the eight dimensions may be independent. For instance, this imposition is revealed explicitly when we factor the coordinate as

$$\underline{\underline{\chi}} = w_0\left(\hat{r} + \hat{i}\ Ht / w_0\right)\left(\hat{\gamma}_0 + \hat{\gamma}_1 w_1 / w_0 + \hat{\gamma}_2 w_2 / w_0 + \hat{\gamma}_3 w_3 / w_0\right). \quad (9.11)$$

Finally we note that there will be more than one arrangement of coordinate axes that exhibits the correct set of symmetries. For instance, the spatial-parity and temporal-sense symmetry-inverse subspaces of a valid subspace are equally valid. Hence, these alternate symmetry-eigenvalue arrangements are equally a part of the derived overall coordinate space. These extra dominating sub-networks are incorporated mathematically as different branches of a general pan-dimensional coordinate space. This pan-dimensional coordinate space can be drawn holistically by using branch cuts to separate the branches in precisely the same manner as branch cuts are used in the mathematical field of complex analysis.

Thus a branch designator completes the mathematical specification of the spatial coordinate. However, we request that the branch designators are coherently expressed. Hence, the same symmetry rules apply and these rules require that the branch designators exhibit the same SU(2) and Q_8 spatial symmetries. Thus the picture for coordinate space that emerges in eventual reality is one in which there are eight branches with a branch designator expressing both quaternion and mathematical-complex number symmetries. We shall come to associate charge conjugation with this final symmetry set.

Thus the coordinate representation for continuum coordinate space takes the form of interwoven subspaces and the full-space components can be written in the form

$$\underline{\underline{\Xi}} = \begin{pmatrix} +_T \\ -_T \end{pmatrix}\begin{pmatrix} \hat{r} \\ \hat{i} \end{pmatrix}\begin{pmatrix} +_P \\ -_P \end{pmatrix}\begin{pmatrix} \hat{\gamma}_0 \\ \hat{\gamma}_1 \\ \hat{\gamma}_2 \\ \hat{\gamma}_3 \end{pmatrix}\begin{pmatrix} +_C \\ -_C \end{pmatrix}\begin{pmatrix} \hat{c}_0 \\ \hat{c}_1 \\ \hat{c}_2 \\ \hat{c}_3 \end{pmatrix}\begin{pmatrix} +_B \\ -_B \end{pmatrix}\begin{pmatrix} \hat{b}_0 \\ \hat{b}_1 \end{pmatrix} a \quad (9.12)$$

where we take one item from each column bracket. Here, the $\hat{c}$ are quaternion-type designators and the $\hat{b}$ are mathematical-complex designators for the branch structure of space and a is a positive mathematical-real number. All terms in this expression belong to independent product groups. The order of writing products of the independent product groups is unimportant. Once again, a consistent use of mathematical signs requires that we are able to factor the branch structure (as well as the dimensions on each branch). This factoring makes it natural to consider together the forward-in-time branch set and the backward-in-time branch set.

Sometimes, it will be convenient to denote the pair of *directional* designators as $\hat{x}_m$ and to denote the pair of *branch* designators as $\hat{y}_n$. Then the coordinate components are abbreviated as

$$\underline{\underline{\Xi}} = \Xi_{mn}\hat{x}_m\hat{y}_n \qquad m,n \in (0..7). \quad (9.13)$$

The presented structure for continuum coordinate space is a key prediction of eventual reality. The result represents a clear differentiation of the theory of eventual reality from other current theories. The analysis implies that the fundamental universe is an over-abundant jumble of superimposed sub-networks that will appear to be like a space with eight-branches, each with four complex-real and four complex-imaginary dimensions.

We will show that this representation of spatial symmetries can explain several features and properties of the known fundamental particles. Furthermore this picture clearly agrees with an observable space possessing four space-time dimensions. Eventual reality does not include any additional curled-up minor dimensions. Thus all of the explicit dimensions of this space, other than the branch structure, have been directly observed.

9.3.1 Objections to a Complex Quaternion Space

Attempts to associate quaternion mathematics with geometrical space are not new. However, there are several seemingly good reasons why such spaces should be rejected.

The first reason is associated with the particular algebraic symmetries proposed. Different algebrae create geometries with different instructions for evaluating the length of a four-vector. Attention is paid to the signs of Cartesian contributions to the square of the rotationally invariant four-vector length. The pattern of signs defines the algebraic signature of a space[18]. Thus the sequence of signs in the Klein-Gordon equation may be used to display the Lorentzian signature of a flat relativistic space.

$$\left(m_0c^2\right)^2 = \varepsilon^2 - (p_xc)^2 - (p_yc)^2 - (p_zc)^2 \qquad (9.14)$$

$$\text{Lorentzian Signature} = (+,-,-,-). \qquad (9.15)$$

The sign convention is unimportant hence using different mathematical language, with one contrary numeric sign, this signature is attributed to a space with an inertial index of one. Now, the invariant scalar length of a quaternion number is obtained by multiplying the quaternion by its quaternion *conjugate* (see equation 3.12) yielding the

$$\text{Quaternion Signature} = (+,+,+,+) \qquad (9.16)$$

ie a space with an inertial index of zero. Thus there is a clear disagreement between the experimentally confirmed Lorentzian signature of observed space and the seemingly more rational signature of a quaternion space. Consequently the space that we directly access by measurement and observation cannot be a simple mathematical-complex quaternion space. However, this seemingly unsatisfactory signature does not signify that a mathematical-complex quaternion space cannot be the fundamental basic space of reality. Indeed, we shall show that in order to build consistent local ordering and regularity, the mathematical-complex quaternion coordinate space is logically transformed into a different space that indeed satisfies the observed signature.

Another reason for rejecting this derived complex-quaternion space is philosophical. The generated time variable is unbounded; hence, time must run from minus infinity to positive infinity. Whilst proposing a 'start' to time may pose it own philosophical dilemmas, some special cosmic times are required if the universe (as a whole) is to evolve. Otherwise, if the universe existed forever, then it is to be expected that no time could be more significant than any other time and some form of Hoyle's steady-state universe should prevail. In these circumstances, we would have doubtful reason to be here 'now' and doubtful reason not to have been over-run by an *infinite* number of 'civilizations' originating in countless other galaxies.

The accepted key to settling this conundrum is to establish something special and relevant about the present now or the not too distant past. Thus we may ask whether there are predictable special times of significance for large distinguishable regions of the universe. As an example, an imagined universe that went through cycles of expansions/bangs and contractions/crunches would have a series of special times yet the universe could have unbounded time. However, it will be explained that a cyclical resolution is not the problem's solution that is envisaged by eventual reality. In the storey of reality holding coordinate space, there is no requirement for a start to time. The temporal argument applies to the reality observed in our apparently matter-occupied causal space. We shall observe that experienced space is derivable by two processes. The first process involves a coordinate transformation. Secondly the temporal way, in which information is registered in this representation of space, requires a unit system that merges forwards and backwards time-coordinate variables into a single observable-time variable. Thus measured time from the big bang is rendered always positive. Hence, the origin of these coordinates, identified as a special location at which all of the branches meet, is shown to correspond to *the* 'big bang'. In our space the future is unbounded, the past is bounded.

There can be other reasons why we don't recognize this coordinate space as describing our observed universe. The relation between the coordinate values eg

$$\frac{w_0}{t} = c\frac{w_1}{x} \qquad (9.17)$$

(which is equivalent to $p_1 = mx_1 / t$) does not fit mechanical expectations. Of course this difficulty could be a simple misinterpretation of the significance of the coordinate axes. However, we shall show that our expectations are appropriate to a different viewpoint in our developing hierarchical scheme of viewpoints. Coordinate space is a foundation space. Our reality involves fluctuations in this foundation coordinate space. Thus as our hierarchical interpretation of the universe builds, space occupied by matter will be transformed into a more recognizable form.

9.4 The Content of Space

We have associated a special package of dual symmetries with the coherent structure that leads to demonstrative regularity in the universe. We shall now observe how these symmetries also describe structures in space that eventually lead to the descriptions of the content of space.

Now, we cannot ignore all the pointer references that don't contribute to the dominating sub-network of coordinate space. Many of the events that are assigned to coordinate space, have additional stray pointer references to events that are 'otherwhere' (ie possibly an unsuitable elsewhere in coordinate space but also more often nowhere in coordinate space). Each stray pointer link, belonging to an event in coordinate space and pointing nowhere, is part of a non-space networked environment for this event. However, just like the full cosmic network, coherent dominating features can be attributed to this dishevelled non-space environment. From the perspective of coordinate space, these stray links resemble a stubble-like growth in coordinate space – a growth that points into otherwhere land. We use mathematical operations in quantitative space when we seek some combined resultant coherence out of all the extra-spatial pointing since we coarse-grain average over spatial regions and even combine results across branches. Effectively the symmetry of quantitative space imposes itself upon our view of the qualitative aethereal stubble sprouting into otherwhere land. Hence although this stubble does not possess numeric extension (as recognized by coordinate space) it still displays the ubiquitous SU(2) and Q_8 spatial symmetries.

Moreover, in a universe in which everything happens (premise 1) we should not presume that these dual symmetries are *singly* expressed in the network stubble. Thus the dominating structure of stubble will possess a repeated sequence of the packaged dual symmetries, the ground level of this hierarchy is the coordinate-space. The higher levels can be taken to form non-extended qualitative multilevel field embedded in the ground-level coordinate space.

Now, this inherently immeasurable all-pervasive set of higher-level structures is whimsically (even provocatively) called the aether[19]. The aether is not a matter field per se. Indeed the aether has intrinsically qualitative features that are best described as a universal language. A consideration of this language will be made in chapter 12; this study shows that the disciplines of logic and mathematics are requirements of eventual reality. Of course, in a self-consistent loop, mathematics and logic were needed to deduce the nature of the universe from the axiomatically defined events.

Every qualitative and quantitative thing, of which we can possibly be aware, is a consequence of the structure of our aethereal universe. However, as the existence of events is not testable, those disbelieving of rational philosophical arguments may prefer to take a tumultuous universe with repeating (SU(2) and Q_8) symmetries as the fundamental description of eventual reality. All the empirical scientific evidence that we are ever likely to gather will be in support of this geometrical basis for eventual reality rather than the fundamental event-based structure of the apeiron, that we have just described.

We have constructed a flat space-time. Thus the aether contains all accessible knowledge about what is 'in' this coordinate space. The multi-branched matter field is derived from knowledge of the aether as a mathematical-complex quaternion probability density on each branch. We shall return to a discussion on the aether in chapter 12. In addition, details of the information embedded in the distortions of space will be discussed in part 3.

Having uncovered the role of the aether we may now return to explaining the branches of space. We have revealed the organizational structure of the spatial branches in two distinct ways. Therefore it is useful to observe how these ways are related ie how the symmetries connecting the branches relate to the division-ring elements that we have attached to the branches. We have noted that the spatial structure of each branch is stationary and these individual structures are understood as the superposed sub-networks assembled with different coordinate symmetries. However, in addition to the spaces on each branch, the set of branches must itself be stationary. To describe the ensuing symmetries we introduce a set of operators. Hence we define an operator P_B which acts on one branch designator to reveal another designator labelling the paired branch with the opposite parity. Note that this operator only acts on the branch designator; it does not operate on the spatial coordinates on a branch. Likewise we define an operator T_B which converts one branch designator into another designator denoting the paired reversed-time-sense branch. We also generate a distinct branch by applying another symmetry operator $T_B P_B$.

By introducing this set of non-identity operators, we have the option to add a new mathematical sign. This freedom becomes a necessity by virtue of premise 1[20]. Hence

$$P_B^2 = T_B^2 = (T_B P_B)^2 = (-)_C \,. \qquad (9.18)$$

In other words, the relativities within a space are unaffected by a double reflection; however, the doubly reflected branch is distinguishable from the original. Consequently these operators anticommute eg

$$(T_B P_B)(T_B P_B) = (T_B P_B)(-)_C (P_B T_B) = T_B^2 = (-)_C \qquad (9.19)$$

$$(T_B P_B) P_B = T_B P_B^2 = (-)_C T_B \qquad (9.20)$$

$$P_B(T_B P_B) = P_B(-)_C P_B T_B = T_B \,. \tag{9.21}$$

Therefore it follows that these operators are associable with the quaternion branch designators as

$$P_B \,\hat{c}_0 = \hat{c}_1 \tag{9.22}$$

$$T_B P_B \,\hat{c}_0 = \hat{c}_2 \tag{9.23}$$

$$T_B \,\hat{c}_0 = \hat{c}_3 \tag{9.24}$$

However, these quaternion branches do not give the complete picture. Our space is also pervaded by the aether. The aether is allowed its own mathematical sign and hence the aether provides a lone reflection symmetry generated by operator, B_B with

$$B_B^2 = (-)_B \,. \tag{9.25}$$

Hence this symmetry corresponds to a paired mathematical-complex SU(2) branch symmetry which becomes factored out of the branch structure.

$$B_B \hat{b}_0 = \hat{b}_1 \tag{9.26}$$

In the next chapter, it will be argued that the procedure of optimizing a local representation for the aethereal matter field can yield a familiar quantitative science in a space with particle fields satisfying the quantum-mechanical equation describing state propagation. A particle-field breakdown of the aethereal field is not surprising; the equivalence of a many-particle field with a continuum aethereal field is simply the equivalence noted by a (generalized) density-functional theory.

Notes

[1] The terms macroscopic and microscopic are used to refer to large-scale and small-scale views of the same system. However, as we change our descriptions of reality what we previously called macroscopic in one context might be referred to as microscopic in another context. For instance, from a material perspective, we may say that classical mechanics provides a macroscopic picture and quantum mechanics a microscopic picture. However, from an aethereal perspective, we will say that quantum mechanics provides a macroscopic description.

[2] The generalized version (implied here) of Fermat's principle was described in section 6.3.3. The role of the theorem, also referred to as the first principle of the universe, is explicitly invoked in proposition 1.

[3] For a discussion on these worlds see Popper (1974) pp180-7. The ancient Greek philosophers made the first proposals of the existence of a figurative world. Wondering over the eternal truth of mathematics and logic, they concluded that these fields of study were closer to the divine than the physical world and these fields actually must exist in another style of world. Thus in an act of self-aggrandisement,

these philosophers proclaimed that those who understood mathematics and logic and who hankered after an ultimate truth were closer to God than the plebs.

[4] Heisenberg (1958) pp 44-58 provides an excellent exposition on the nature of quantum measurement and measurement's relation to quantum reality. He argues that the only psychologically satisfying reality is one based upon the realism contained within classical mechanics.

[5] The change of terminology from *premise* to *proposition* is used to convey an implication that the propositions are, in a sense, more negotiable and perhaps more anthropologically based. The propositions are suited to the way we like to describe science. The premises are less tainted.

[6] An explanation of the scale is provided in section 17.4.1.

[7] This terminology will be explained in the chapter. Eventual reality refers to all storeys in the edifice of reality. The storey immediately below the principles of quantum mechanics houses the foundation space known as coordinate space.

[8] In an infinitely dimensioned vector space we would need an infinite set of numbers just to identify a single point. However, similarly we would need an infinite set of numbers to define a plane! Furthermore the distance between almost every pair of points would be infinite.

[9] Here we are creating a quantitative space from the intangible qualitative apeiron. Later, in section 13.2, we will justify the existence of the apeiron, itself.

[10] Consistency and completeness allow us to claim a symmetry that may be or may not be trivial.

[11] Adler (1958) comprehensively describes the properties and associations of the line of mathematical-real numbers. This work demonstrates how mathematicians view numbers.

[12] Affine conditions are defined in table 15.1.

[13] For details and references to division-ring algebrae over the mathematical-real numbers see Iyanaga & Kawada (1977) §31B.

[14] Devlin (1988) pp 73-74 account of mathematics shows how to view quaternion numbers within the range of forms of mathematics.

[15] Individually the SU(2) and Q_8 groups form division rings over the mathematical-real numbers. However, there is a problem when we form a combined group.

[16] See discussion in section 3.2.

[17] For a fuller explanation of quaternion algebra see section 3.3.1.

[18] For comments on the invariance of the signature of space see Penrose (2004) §11.2 and Weyl (1952) §16.

[19] Traditionally the aether has been an accepted yet unproven entity designed to plug a void in the conceptual picture. The term was introduced by Aristotle to describe an eternal spiritual celestial (ie beyond the Moon) essence that denoted purity and permanence. However, in classical science the aether was the all-pervasive medium in which, it was thought, a wave of light or radiated energy must propagate.

[20] The algebra allows a reverse-parity branch to be formed; therefore it is formed.

Part 2

Chapter 10 : Quantum Matter

Eventual reality proposes to displace quantum mechanics at the forefront of the foundations of knowledge whilst retaining the predictions of quantum mechanics ('the most successful physics theory ever') in some restricted 'macroscopic' range[1] of applicability. It is possible to regard quantum mechanics as the progenitor of much of physical theory. Through quantum mechanics, the theory of eventual reality can lead to classical mechanics, statistical mechanics, electrodynamics, optics, etc.

If we accept that quantum mechanics can only be a statistical theory, then we should not be surprised that we can invent a statistical model that reduces to quantum mechanics. Indeed a correspondence, between the path-integral formulation of quantum mechanics and statistical mechanics, is already well known. Consequently a subsumption of quantum mechanics by eventual reality[2] is to be reasonably expected and this subsumption cannot be substantially relied upon to justify the adoption of the specific nature of eventual reality over other as yet unformulated statistical theories. In other words, being able to derive general quantum principles from the premises of eventual reality is support for a statistical hypothesis rather than support for eventual reality per se.

Nevertheless the derivation, of quantum mechanics from eventual reality, is not straightforward and there are many features of the result that are characteristic of the details of the hypothesis of eventual reality. These specific characteristic features include the hidden symmetries displayed by the spatial branches and the aethereal field, and a clear requirement for the introduction of particular mathematical number forms. The explanations of quantum measurement and of measurement's relation to the principles of quantum mechanics are also further support for the structural details of the hypothesis of eventual reality.

In this chapter we shall investigate how quantum formulations can be obtained as the macroscopic form of eventual reality and thus we shed light upon the limits of quantum applicability and we shed light upon the properties of physical space and matter. This analysis applies whenever there is no consistently applied intent. Such issues associated with intent will not be fully explored in this work. However, it is understandable that random statistics will not apply whenever conscious actions[3] are permitted.

Clearly the descriptions of quantum mechanics and eventual reality are very different. In eventual reality, aethereal fields in a multi-branched space express the material details of the universe, whereas, quantum mechanics was constructed to describe the mathematical-real scalar results of localized measurement within a framework based upon our classical notions supported by material monism. Thus in quantum mechanics, matter is visualized as a non-localized state that has cognizance of an indefinite number of interacting single-particle fields.

Recall that, there are many essentially equivalent representations of quantum mechanics. For instance, there are the different 'pictures' (associated with the choice of a system of basis states) there are representations classed as first and second quantizations (associated with a choice of fully representing system dynamics by states or operators) and there are ways of expressing the dynamics in differential and integral forms (often derived from Hamiltonian and Lagrangian formulations, respectively), etc. Consequently the analysis that follows is only one example of many analyses that could reveal the transition from eventual reality to the union of approaches known as quantum mechanics. In practice but not in principle, a derivation of any one of these representations is effectively a derivation of them all.

We find ourselves in an unusual (and fortunate) situation as we derive quantum mechanics – we already know the answer. This pre-knowledge can help us interpret (and even guide) the derivation. However, even though we deliberately set out to derive quantum mechanics we are not forcing the answer. There are many ways to describe reality. Here we are not interested in a bland description such as the evolution of the mean matter density of the universe. We are interested in a highly detailed description with strong spatial discrimination. Thus we might discuss the progress of the development by referring to quantum mechanics or even classical mechanics but what we are actually asking for is a rich description of the aethereal field as matter in space.

The analysis will draw, in part, upon many conventional approaches including the Lagrangian formulation of classical mechanics, the theory of collective coordinates and Fourier and Hilbert analyses. Evidently the transition from eventual reality to quantum mechanics needs several steps.

10.1 Macroscopic Averaging

In the ensemble that represents the universe single events have little individual weighting. Thus our physical reality is based upon the consistent patterns displayed collectively by very many events. Before we embark upon a (collective coordinate) procedure to reveal the structures within these patterns, we shall look at the impact of macroscopic spatial averaging.

The full extent of the accessible knowledge available in the aether is described by a multi-level field, $\underline{\underline{K}}\left(\underline{\underline{\Xi}}\right)$, pervading coordinate space. However, this field contains more information than is pertinent to just the physical world. Here, $\underline{\underline{\Xi}}$ is an eight coordinate and eight-branched space. Note that where appropriate we shall assign *subscript* coordinate and branch indices to ground-level space and assign superscript indices to the next higher-level aethereal space. In part 2 of this document we shall not use the subscript and superscript conventions of tensors. To make this unconventional[4] notation more obvious, commas are used to separate indices.

By definition, the inanimate/inert physical world is devoid of coherent mental-style activity. This world is not governed by design or purpose but happenstance. The descriptors of this physical world are order and regularity (as described by quantitative information). Hence, the relevant quantitative information content will be represented by another mathematical-complex quaternion field, $\underline{\underline{\Phi}}^{m_u,n_u}\left(\underline{\underline{\Xi}}\right)$, in coordinate space. This new function describes the degree of regularity and order in the aethereal field, $\underline{\underline{K}}^{m_u,n_u}\left(\underline{\underline{\Xi}}\right)$, thus $\underline{\underline{\Phi}}^{m_u,n_u}\left(\underline{\underline{\Xi}}\right)$ is an entropy parameter (chosen to be a monotonic function of entropy).

It is convenient to work with an entropy parameter that resembles a probability density. Multiplying such parameters across the branches (at the same $\hat{x}$-space location) is not unlike adding their entropies. Care is needed with such manipulations; a probability density is not generally proportional to the exponential of the negative of the entropy density since quaternion numbers are non-commuting. The product of entropy parameters describes the likelihood of a local value being consistently realized over branches. Note that at each location on each branch, the number of choices (linking all branches) takes a numerical value that is a mathematical-complex quaternion not a mathematical-real scalar. Thus the probability densities on individual branches of space are only exceptionally mathematical-real positive-definitive quantities.

We can define macroscopic field values for the metaphysical (ie multi-branched) spatial coordinate and the meta-fields by averaging over a small region of coordinate space around each physical coordinate spot as

$$\underline{\underline{\xi}} = \left\langle \underline{\underline{\chi}} \right\rangle_{\chi \in \Omega} \qquad (10.1)$$

for each lowest level branch. Thus functions of the branchless $\underline{\underline{\xi}}$ may be provided with an additional index to denote the coordinate branch:

$$\underline{\underline{S}}_{l}^{m_u,n_u}\left(\underline{\underline{\xi}}\right) = \left\langle \underline{\underline{\Phi}}^{m_u,n_u}\left(\underline{\underline{\chi}}\right) \right\rangle_{\chi \in \Omega} \qquad (10.2)$$

and similarly

$$\underline{\underline{\Psi}}_{l}^{m_u,n_u}\left(\underline{\underline{\xi}}\right) = \left\langle \underline{\underline{K}}^{m_u,n_u}\left(\underline{\underline{\chi}}\right) \right\rangle_{\chi \in \Omega} . \qquad (10.3)$$

Here the bra (<) and ket (>) parentheses symbolize an average taken within the coordinate-space 8-volume, Ω. The size of this volume is to be very much larger than the unit hyper-cube volume based upon the scale length attributed to an event. If we displace this volume by a few discrete events, most of the events within the volume will remain in the volume. Nevertheless the displacement will cause many events to leave or enter the volume.

Therefore, in the absence of large-scale coherence over the surface of region, Ω, we expect the average value of the field to vary smoothly with the displacement. Thus one aspect of the transition from a microscopic description to a macroscopic description involves a transition from potentially discrete chaotic field elements to a smooth analytic continuum field. Analyticity of the physical equations and continuity of variables and are *not* intrinsic to the underlying microscopic structure, analyticity and continuity are a reflection of a large-scale macroscopic resolution.

We expect that a macroscopic theory constructed from microscopic eventual reality may have only a limited access to the underlying microscopic reality and such a theory may involve the creation of both entirely new and evolved physical entities. In electromagnetism, the macroscopic descriptions introduce the magnetic and electric displacement fields; in statistical mechanics, the macroscopic descriptions introduce the continuum thermodynamic fields of pressure, density, free energies, etc. Similarly the detail of the quantum-mechanical worldview is quite different from the detail of the event-filled worldview of eventual reality. For instance, quantum mechanics is critically influenced by both the mathematical-real scalar nature of measurement and by discrete localized constructions known as particles. Hence, it is appropriate to consider how new concepts are introduced and how they might relate to a local mathematical-real scalar viewpoint.

The quantum picture is one of fields of particles. Thus in order to create a quantum style picture of the analytic dependence of $\underline{\underline{\Phi}}_l^{m_u,n_u}$ upon $\underline{\underline{\xi}}$, we will need to create intermediary fields such that $\underline{\underline{\Phi}}_l^{m_u,n_u}$ is functionally dependent upon these intermediate fields which, in turn, are functionally dependent upon $\underline{\underline{\xi}}$.

In conventional mathematics, a general field can be segregated into individual orthogonal subcomponents by making an expansion in terms of a complete orthogonal set of functions. A simple example of such an expansion is provided by the Fourier transformation. Each term in the expansion can be viewed as a discrete excitation and a subcomponent of the total field. We shall aim to generate subcomponent fields in a similar manner using a version of the technique known as collective coordinates[5]. This set of collective component fields will sample aspects of the scale-dependent coherent behaviours of the aether.

Thus for each element of the matrix, $\underline{\underline{\Phi}}_l^{m_u,n_u}$, we wish to construct a set of spatial collective field elements of Fourier style. To perform the appropriate transformations, we shall identify an isolated region of coordinate space and subdivide this isolated region into a very large number of discrete equal-size cells. We will retain the scale size of these discrete cells as the determiner of the appropriate number of representational variables/particles (both real and virtual) in the description of the system.

Unfortunately the exponential function for quaternions does not follow the familiar useful properties of the exponentials of regular (integral-domain) numbers. Critically we cannot construct a reciprocal coordinate transformation in an analytic fashion. Thus a standard Fourier expansion cannot be performed in quaternion algebra.

Therefore there are two ways in which we could proceed. We could seek an alternative to the reciprocal transformation method, ie an alternative method of generating a rich description involving a sequence of subcomponents. Alternatively we could seek a new algebra that will enable us to achieve the desired reciprocal transformation. It is very likely that both approaches would be successful. The latter will generate quantum mechanics; the former probably needs a new mathematical structures. Here we will pursue the latter method.

The issue of number forms is at the heart of one essential different between quantum mechanics and eventual reality. The eventual-reality description of coordinate space is analytical in mathematical-complex quaternion numbers. However, (classical) scientific measurements yield localized results composed of mathematical-real scalar numbers. Apparently quantum mechanics is forced to be a hybrid link between these dichotomies.

10.2 Number Forms

To facilitate an analytical breakdown of the full aethereal field, we introduce a transitional number form that can be used to extract mathematical-real quaternion-scalar elements from the underlying mathematical-complex quaternion structures and thereby avail ourselves of the correspondences of reciprocal transformations. This transitional number form engineers a link to the accepted analytical quantum reality portrayed by measurement. This new number form is referred to as a quasi-number and is governed by an extension of the mathematical-complex quaternion algebra. I am referring to quasi-numbers as transitional since their use with mathematical-complex quaternions creates yet another algebra, a new algebra which is directly apparent in the quantum realm. Hence quasi-numbers are introduced to provide a link to this other algebra.

Now, quaternion numbers have both a quaternion-vector and a quaternion-scalar part and thus quaternions describe a more elaborate topology than that topology displayed by standard 3-vectors.

$$(a_1, \underline{v}_1)(a_2, \underline{v}_2) = (a_1 a_2 - \underline{v}_1 . \underline{v}_2, a_1 \underline{v}_2 + a_2 \underline{v}_1 + \underline{v}_1 \times \underline{v}_2) \qquad (10.4)$$

Observe that quaternion algebra contains a vector algebra as an incomplete sub-algebra. We shall explore the viability of introducing a set of additional number entities that can exploit this sub-algebra (ie resemble mathematical-complex vectors) and that can be used as Fourier conjugates to mathematical-complex quaternion coordinates. The elements of this set are invented solely for generating the quasi-scalar product required in reciprocal transforms. Thus the (quasi-scalar) products of quasi-numbers with (mathematical-complex quaternion) coordinates will be required to work effectively with standard mathematical-complex quaternion arithmetic. Consequently, in this section, we shall explore both the quasi-number algebra and their mixed algebra with functions of coordinates.

We start our discussion by considering a single quasi-scalar product. Now, by definition, a quasi-number, $\overline{\overline{q}}$, needs to possess a product with the mathematical-complex quaternion number, $\underline{\underline{\Gamma}}^{(j)}$, as

$$\underline{\underline{\Gamma}}^{(j)}\overline{\overline{q}} = (-)_P\left(\sum_{i=1..3} q_i\Gamma_i^{(j)} + (-)_T\sum_{i=5..7} q_i\Gamma_i^{(j)}\right)\hat{\gamma}_0\hat{r}\hat{e}_F^{(j)} = \overline{\overline{q}}\underline{\underline{\Gamma}}^{(j)}. \quad (10.5)$$

Here, $\overline{\overline{q}}$ is a new mathematical-complex vector-like entity that is deemed to have a special product with a mathematical-complex quaternion vector. $\overline{\overline{q}}$ is not an acceptable element of the mathematical-complex quaternion number system and hence the naming of these vectors as quasi-numbers. Note that the absence of a double underline or double overline accompanying the directional components of $\underline{\underline{\Gamma}}^{(j)}$ and $\overline{\overline{q}}$ (inside the large round brackets in equation 10.5) indicates that these components q_i and $\Gamma_i^{(j)}$ are mathematical-real numbers (ie the q_i and the $\Gamma_i^{(j)}$ terms do not include any designators).

The spatial-dimension indices for $\underline{\underline{\Gamma}}^{(j)}$ and $\overline{\overline{q}}$ run from 0 to 7; 0 denotes the complex-real scalar term, 1 to 3 denote the complex-real vector terms, 4 denotes the complex-imaginary scalar term, 5 to 7 denote the complex-imaginary vector terms. However, the quasi-scalar product does not involve the quaternion-scalar dimensions: time (dimension 4) and specific energy (dimension 0). These conventions are displayed in table 10.1. Thus the Fourier transformation indirectly erases the full quaternion symmetry. Nevertheless all quaternion-vector directions are treated equally.

In keeping with the representation of generalized coordinates in the theory of eventual reality, a set of unit-sized designators, $\hat{e}_F^{(j)}$, has been introduced in equation (10.5). This designator is used to signal that a product involving a quasi-number has been undertaken. The designators are nominally unit sized. This unit-size specification ensures that the magnitude of the product is given by the component magnitudes ie the length of the quasi-scalar product conforms to normal expectation. Thus

$$\left|\hat{e}_F^{(j)}\right| = \hat{e}_I \quad (10.6)$$

where $\hat{e}_I$ is a designator that performs the role of an identity for the set of new designators, $\{\hat{e}_F^{(j)}\}$. Thus $\hat{e}_I$ designates a number belonging to the coordinate-number system of mathematical-complex quaternions ie the designator identity indicates the base number form when the additional numbers spanned by the $\hat{e}_F^{(j)}$ designators are absent.

Table 10.1

Notational conventions that apply to the unit-dimensionless coordinates of the primary spaces. Mathematical-complex quaternion numbers occupy coordinate space and quasi-numbers occupy quasi-space.

	Designator	Coordinate Space	Quasi-Space
0	$\hat{r}\hat{\gamma}_0$	w_0 or $\varepsilon / m_0 c^2$	
1	$\hat{r}\hat{\gamma}_1$	w_1 or $p_1 / m_0 c$	HX_1 / c
2	$\hat{r}\hat{\gamma}_2$	w_2 or $p_2 / m_0 c$	HX_2 / c
3	$\hat{r}\hat{\gamma}_3$	w_3 or $p_3 / m_0 c$	HX_3 / c
4	$\hat{i}\hat{\gamma}_0$	Ht	
5	$\hat{i}\hat{\gamma}_1$	Hx_1 / c	W_1 or $P_1 / m_0 c$
6	$\hat{i}\hat{\gamma}_2$	Hx_2 / c	W_2 or $P_2 / m_0 c$
7	$\hat{i}\hat{\gamma}_3$	Hx_3 / c	W_3 or $P_3 / m_0 c$

The order of multiplication follows from the arbitrary convention for the order of multiplying branches.

$$\hat{e}_I \hat{e}_F^{(j)} = \hat{e}_F^{(j)} \hat{e}_I = \hat{e}_F^{(j)} \text{ and } \hat{e}_I^2 = \hat{e}_I . \qquad (10.7)$$

Each distinguishable quasi-scalar product is an affront to mathematical-complex quaternion arithmetic and hence a separate designator, $\hat{e}_F^{(j)}$, is introduced for every distinct quasi-scalar product. Without a compelling reason to make these designators the same, we have to allow each of these new designators to be different. We will uncover how members of this set of designators are to be related in section 10.2.2.

Now, the sets of designators naturally form a hierarchy of different number styles. Thus if we only work with mathematical-real numbers, we would not need to use the designator, $\hat{r}$. If we work with mathematical-complex numbers but not quaternions, we would not need to use the designator, $\hat{\gamma}_0$. Likewise, if we only worked with mathematical-complex quaternions, we would not need to use the designator, $\hat{e}_I$. In this work, we shall not extend the algebraic representation of numbers beyond the arithmetic system containing designators, $\hat{e}_F^{(j)}$.

Next, we shall investigate what further properties of quasi-numbers are needed to provide a workable extension to our mathematical-complex quaternion coordinate algebra. This extension algebra will attempt to retain as many unextended properties of numbers as possible. For notational convenience, the order of precedence of multiplications (in the absence of explicit brackets) will be taken in the order: quasi-number with quasi-number, then quasi-number with mathematical-complex quaternion and finally mathematical-complex quaternion with mathematical-complex quaternion.

For all expressions *linear* in $\overline{\overline{q}}$, we may view $\overline{\overline{q}}$ as behaving like a mathematical-complex quaternion with a zero quaternion-scalar component. However, after we perform a quasi-scalar product, it will be necessary to take the mathematical-real quaternion-scalar part of the answer.

Explicitly displaying mathematical-complex quaternion numbers as a mathematical-complex binary column vector formed from a comma separated scalar/vector quaternion pair, the quasi-scalar product becomes

$$\overline{\overline{q}}\underline{\underline{\Gamma}}^{(j)} = Re\, Sc\begin{pmatrix} 0,\underline{u}_1 \\ 0,\underline{v}_1 \end{pmatrix}\begin{pmatrix} a_2,\underline{u}_2 \\ b_2,\underline{v}_2 \end{pmatrix}\hat{e}_F^{(j)}$$

$$= Re\, Sc\begin{pmatrix} (-)_P(\underline{u}_1.\underline{u}_2 + (-)_T\underline{v}_1.\underline{v}_2), a_2\underline{u}_1 + \underline{u}_1\times\underline{u}_2 + (-)_T(b_2\underline{v}_1 + \underline{v}_1\times\underline{v}_2) \\ (-)_P(\underline{v}_1\,\underline{u}_2 + (-)_T\underline{u}_1.\underline{v}_2), a_2\underline{u}_1 + \underline{u}_1\times\underline{v}_2 + b_2\underline{v}_1 + \underline{v}_1\times\underline{u}_2 \end{pmatrix}\hat{e}_F^{(j)}$$

$$= (-)_P\left(\underline{u}_1.\underline{u}_2 + (-)_T\,\underline{v}_1.\underline{v}_2\right)\hat{\gamma}_0\hat{r}\hat{e}_F^{(j)}\,. \qquad (10.8)$$

Prior to performing the non-analytical mathematical-real-part and quaternion-scalar part operations, the mathematical rules of quaternion algebra carry over into this algebra. The quaternion-scalar part, Sc, and the mathematical-real part, Re, operations are performed last. Thus the $\hat{e}_F^{(j)}$ give us a way of analytically handling what would non-analytic in the other algebrae alone.

Consequently the quasi-scalar products satisfy the distributive laws

$$\overline{\overline{q}}(\underline{\underline{\Gamma}}^{(j_1)} + \underline{\underline{\Gamma}}^{(j_2)}) = \overline{\overline{q}}\underline{\underline{\Gamma}}^{(j_1)} + \overline{\overline{q}}\underline{\underline{\Gamma}}^{(j_2)}$$

and $\qquad (10.9)$

$$(\underline{\underline{\Gamma}}^{(j_1)} + \underline{\underline{\Gamma}}^{(j_2)})\overline{\overline{q}} = \underline{\underline{\Gamma}}^{(j_1)}\overline{\overline{q}} + \underline{\underline{\Gamma}}^{(j_2)}\overline{\overline{q}}\,.$$

However, unconventionally the product of a quasi-number with a quaternion scalar, $\underline{\underline{S}}$, is zero.

$$\overline{\overline{q}}\underline{\underline{S}} = 0 \qquad (10.10)$$

In a multiplication operation, the quasi-number acts only on the six quaternion-vector components. Nevertheless, as we are about to discuss, if we multiply the quasi number with an integral-domain complex number we do scale the quasi-number's magnitude. Hence, it is essential to explicitly show (when applicable) the designator $\hat{\gamma}_0$.

Now, we shall let quasi-numbers form a module over the mathematical-complex numbers. Thus if we represent a quasi-number, $\overline{\overline{q}}$, by

$$\begin{pmatrix} 0, \underline{u} \\ 0, \underline{v} \end{pmatrix}$$

then $C\overline{\overline{q}}$ is represented by

$$\begin{pmatrix} 0, C_R \underline{u} + (-)_T C_I \underline{v} \\ 0, C_R \underline{v} + C_I \underline{u} \end{pmatrix} \qquad (10.11)$$

where C is the mathematical-complex number, $\begin{pmatrix} C_R \\ C_I \end{pmatrix}$. Moreover, we have

$$C(\overline{\overline{q}}_1 + \overline{\overline{q}}_2) = C\overline{\overline{q}}_1 + C\overline{\overline{q}}_2 \qquad (10.12)$$

$$1\overline{\overline{q}} = \overline{\overline{q}} \qquad (10.13)$$

and for two mathematical-complex numbers, C_1 and C_2, we have

$$(C_1 + C_2)\overline{\overline{q}} = C_1\overline{\overline{q}} + C_2\overline{\overline{q}} \qquad (10.14)$$

and

$$(C_1 C_2)\overline{\overline{q}} = C_1(C_2\overline{\overline{q}}). \qquad (10.15)$$

Thus it is clear that these quasi-numbers form a left-module[6] over the mathematical-complex numbers, which are formed with designators $\begin{pmatrix} \hat{r} \\ \hat{i} \end{pmatrix}$ and signs $(\pm)_T$.

10.2.1 Quasi-number Extension Algebra

An algebra is a formal set of rules for manipulating the set of its elements. However, quasi-numbers have relevance only in relation to our mathematical-complex quaternion coordinates. Their algebra is not self-contained; quasi-number algebra exists only as an extension of the mathematical-complex quaternion algebra.

Now, let us consider the product of a quasi-number with another quasi-number. In keeping with our portrayal of quasi-numbers as projections of quaternion vectors, the vector-multiplication assignment for the product of quasi-numbers should be expressed as a vector cross product

$$\overset{=}{q}_1\overset{=}{q}_2 = Ve\begin{pmatrix} 0,\underline{u}_1 \\ 0,\underline{v}_1 \end{pmatrix}\begin{pmatrix} 0,\underline{u}_2 \\ 0,\underline{v}_2 \end{pmatrix}\hat{e}_V$$

$$= Ve\begin{pmatrix} (-)_P(\underline{u}_1.\underline{u}_2 + (-)_T\underline{v}_1.\underline{v}_2), \underline{u}_1\times\underline{u}_2 + (-)_T\underline{v}_1\times\underline{v}_2 \\ (-)_P(\underline{u}_1.\underline{v}_2 - \underline{v}_1.\underline{u}_2), \underline{v}_1\times\underline{u}_2 + \underline{u}_1\times\underline{v}_2 \end{pmatrix}\hat{e}_V$$

$$= (\underline{u}_1\times\underline{u}_2 + (-)_T\underline{v}_1\times\underline{v}_2).\underline{\hat{\gamma}}\,\hat{r}\hat{e}_V + (\underline{v}_1\times\underline{u}_2 + \underline{u}_1\times\underline{v}_2).\underline{\hat{\gamma}}\,\hat{i}\hat{e}_V \qquad (10.16)$$

where the vector part operator, Ve, is to be performed after the multiplication and we have used traditional 3-dimensional vector notation with

$$\underline{\hat{\gamma}} = (\hat{\gamma}_1, \hat{\gamma}_2, \hat{\gamma}_3). \qquad (10.17)$$

Here, as before an extra designator, $\hat{e}_V$, is introduced to signify that a product involving a quasi-number has been performed. The product of two quasi-numbers is not a quasi-number, the result is a mathematical-complex quaternion vector multiplied by the designator $\hat{e}_V$. It is easy to show that we could not both have made the product of quasi-numbers a quasi-number and have retained the associative rule of multiplication[7].

Now, the product rule (equation (10.16)) implies that quasi-numbers anti-commute.

$$[\overset{=}{q}_1, \overset{=}{q}_2]_+ = \overset{=}{q}_1\overset{=}{q}_2 + \overset{=}{q}_2\overset{=}{q}_1 = 0. \qquad (10.18)$$

Furthermore the number zero is sharable across the mathematical-complex quaternion and quasi-number symbolic systems ie zero times anything is still zero. This anti-commutation of quasi-numbers applies quite generally including special cases in which they are the same vector, different vector-components of a common vector, a common function of a quasi-number or quasi-numbers computed at different times.

When the quasi-numbers being multiplied are the same quasi-number, we observe that they display their vector character with a zero product

$$\overset{=}{q}^2 = 0. \qquad (10.19)$$

Clearly our definitions readily satisfy a bi-linearity condition

$$[c_1\overset{=}{q}_1 + c_2\overset{=}{q}_2, \overset{=}{q}_3]_+ = c_1[\overset{=}{q}_1, \overset{=}{q}_3]_+ + c_2[\overset{=}{q}_2, \overset{=}{q}_3]_+ \qquad (10.20)$$

for which the bracket product, $[]_+$, of the algebra is the anti-commutator. Furthermore, also trivially, quasi-numbers satisfies the corresponding Jacobi identity which specifies that cyclic combinations of triple products vanish

$$\left[\bar{\bar{q}}_1, [\bar{\bar{q}}_2, \bar{\bar{q}}_3]_+\right]_+ + \left[\bar{\bar{q}}_2, [\bar{\bar{q}}_3, \bar{\bar{q}}_1]_+\right]_+ + \left[\bar{\bar{q}}_3, [\bar{\bar{q}}_1, \bar{\bar{q}}_2]_+\right]_+ = 0. \qquad (10.21)$$

Thus, in their own right, quasi-numbers possess a complex Lie algebra[8] with this bracket product.

Next, let us look at the form of the extension algebra under regular multiplication. We wish to retain as many of the algebraic symmetries that are displayed by (mathematical-complex quaternion) coordinates as are consistent with our operational impositions. Thus we shall require that quasi-numbers satisfy an associative law of multiplication. Consider the triple product

$$\bar{\bar{q}}_1\bar{\bar{q}}_2\bar{\bar{q}}_3 = \bar{\bar{q}}_1(\bar{\bar{q}}_2\bar{\bar{q}}_3) = (\bar{\bar{q}}_1\bar{\bar{q}}_2)\bar{\bar{q}}_3. \qquad (10.22)$$

The first product that is performed yields a mathematical-complex quaternion times the designator $\hat{e}_V$. We shall use the following shorthand notation for this product

$$\bar{\bar{q}}_i\bar{\bar{q}}_j = \underline{\underline{\Gamma}}_{i/j}\hat{e}_V. \qquad (10.23)$$

The remaining operation of the triple product involves taking the quasi-scalar product of a quaternion and quasi-number.

$$\bar{\bar{q}}_1\bar{\bar{q}}_2\bar{\bar{q}}_3 = \bar{\bar{q}}_1\underline{\underline{\Gamma}}_{2/3}\hat{e}_V = \underline{\underline{\Gamma}}_{1/2}\bar{\bar{q}}_3\hat{e}_V \qquad (10.24)$$

Thus, in the language of vector analysis, the first product taken is a vector product and the second product taken is a scalar product. The argument of the result is simply the volume of the parallelepiped formed by the three corresponding vectors. This argument is multiplied by designators $\hat{\gamma}_0$, $\hat{r}$, $\hat{e}_V$ and $\hat{e}_F$. The sequential order in which the two multiplications (eg between quasi-numbers with indices 1 and 2 and then with 3 or between 2 and 3 and then with 1) are performed is unimportant. Of course the first multiplication performed always corresponds to the vector product.

However, it is necessary to consider higher-order products. The associative law for quadruple products imposes a further constraint on the extension algebra. By performing the multiplications in different sequential orders, the quadruple product can be computed in numerous ways.

$$\bar{\bar{q}}_1\bar{\bar{q}}_2\bar{\bar{q}}_3\bar{\bar{q}}_4 = \bar{\bar{q}}_1\left(a_{2/3/4}\hat{\gamma}_0\hat{r}\hat{e}_V\hat{e}_F\right) = \left(a_{1/2/3}\hat{\gamma}_0\hat{r}\hat{e}_V\hat{e}_F\right)\bar{\bar{q}}_4 = \hat{e}_V^2\underline{\underline{\Gamma}}_{1/2}\underline{\underline{\Gamma}}_{3/4} \qquad (10.25)$$

where we have represented the triple products as

$$\overset{=}{q}_i \overset{=}{q}_j \overset{=}{q}_k = a_{i/j/k} \hat{\gamma}_0 \hat{r} \hat{e}_V \hat{e}_F \cdot$$

Potentially these alternative expressions give somewhat different results. The triple product of quasi-numbers is proportional to a mathematical-complex quaternion scalar number. However, mathematical-complex quaternion scalars are orthogonal to all quasi-numbers. Thus the first two expressions for the quadruple product vanish, whereas, the last expression takes the form of a mathematical-complex quaternion number times $\hat{e}_V^2$. Thus, to reconcile the different ways of computing the quadruple product, we impose the condition that

$$\hat{e}_V^2 = 0 \cdot \qquad (10.26)$$

Hence we assert that the quadruple product of quasi-numbers, along with all higher-order products of quasi-numbers, must be zero. In addition, we know that any pair of quasi-numbers must anticommute. Thus second-order polynomial terms formed from quasi-numbers are ignorable in all expressions that are *symmetric* in arrangements of quasi-numbers eg

$$\overset{=}{q}_1 \overset{=}{q}_2 + \overset{=}{q}_2 \overset{=}{q}_1 = Ve\left((-)_P 2 \underline{z}_1 \cdot \underline{z}_2, \underline{z}_1 \times \underline{z}_2 + \underline{z}_2 \times \underline{z}_1\right) \hat{e}_F = 0 \qquad (10.27)$$

where

$$z_i = u_i \hat{r} + v_i \hat{i} \cdot \qquad (10.28)$$

Every symmetric analytical function that involves only quasi-numbers reduces to a linear form.

By permitting the product of two non-zero numbers to equal zero, the mathematics of these quasi-numbers is seen to departure substantially from the requirements of more-familiar (integral-domain) number theory. Thus quasi-numbers, like vectors and matrices, may be treated (when appropriate) as a form of (integral-domain) composite number. Equally the quasi-numbers can be regarded as intrinsic numbers of an extension non-division algebra.

The introduction of these quasi-numbers enables us to naturally extract the quaternion-scalar and quaternion-vector characteristics of the starting quaternion equations. Note that when we deal with quasi-numbers, we use the quasi-number addition and subtraction functions (signs compatible with mathematical-complex numbers).

10.2.2 Grassmann Algebra

Looking at the vector dimensions of two mathematical-complex quaternions, we recall that vectors anticommute

$$(Ve\,\underline{\underline{\Gamma}}^{(i)}) \times (Ve\,\underline{\underline{\Gamma}}^{(j)}) + (Ve\,\underline{\underline{\Gamma}}^{(j)}) \times (Ve\,\underline{\underline{\Gamma}}^{(i)}) = 0 \cdot \qquad (10.29)$$

Here, the anti-commutation can be attributed to the $\hat{\gamma}_\lambda$ designators. Now, the quasi-scalar product only involves these vector dimensions; however, the quasi-scalar product removes the quaternion-vector designators. It might seem reasonable to preserve this anti-symmetry under exchange of indices. Then if

$$\bar{\bar{q}}\,\underline{\underline{\Gamma}}^{(i)}\,\underline{\underline{\Gamma}}^{(j)}\bar{\bar{q}}+\bar{\bar{q}}\,\underline{\underline{\Gamma}}^{(j)}\,\underline{\underline{\Gamma}}^{(i)}\bar{\bar{q}}=0 \qquad (10.30)$$

the anti-symmetry would need to be replicated by the set, $\{\hat{e}_F^{(j)}\}$.

Furthermore, there is a clear mathematical reason why these new designators should anticommute. Quasi-numbers were introduced as an algebraic extension to factorable complex-quaternion numbers; however, to make this extension work we added further designators. Now, the use of designator notation implicitly suggests that these further designators $(\hat{e}_I, \hat{e}_V, \{\hat{e}_F^{(j)}\})$ form the basis of a number system in their own right. The factorable mathematical-complex quaternions describe the framework of the foundation space; the new number system will describe a decomposition of the aether. This decomposition is free to possess its own symmetries.

Historically mathematical-complex numbers were devised as a generalization of mathematical-real numbers; quaternions were devised as a generalization of mathematical-complex numbers. Making the new number system retain as many properties of complex-quaternions as possible, will ensure that the new number system will be seen as a generalization of quaternion numbers.

An axiomatic interpretation of the different algebrae (featuring in eventual reality) is given in the table 10.2. A ring is a non-trivial set of elements with binary operators $(+,\times)$ that forms a group under $+$ and a semi-group under $\times$. We start working with algebrae over the mathematical-real numbers (as defined by the line). Mathematical-real numbers have several more properties than the elements of rings; these properties are surrendered one-by-one during successive generalizations. The generalizations are necessary as the dimensionality of the number system is increased. Clearly a one-dimensional space allows exclusive ordering of numbers. A number system of two or higher dimensions, can only be ranked by one-dimensional comparisons. Without the comparative law, mathematical-real numbers are generalized to mathematical-complex numbers. Here the mathematical-real axis of the complex numbers, are the mathematical-real numbers. Clearly in two dimensions, rotations are commutative. A number system of three or higher dimensions, only has commutative rotations in a two-dimensional plane. Hence, without the commutative product law, mathematical-complex numbers are generalized to quaternion numbers. Here, we may recover mathematical-complex numbers by taking pairs comprised of a quaternion-scalar and a one-dimensional quaternion-vector.

Table 10.2

Axiomatic comparison of the tiered hyper-complex sequence of number systems revealed by eventual reality. Symbols *a*, *b* and *c* represent numbers, *I* represents an identity number. Higher tier algebrae have more dimensions. The Grassmann system used here will be a finitely dimensioned extraction of a system of unlimited dimensionality.

System			Real	Complex	Quarternion	Grassmann
Ring	Rules of Addition	Commutative law: $a+b=b+a$	×	×	×	×
		Associative law: $(a+b)+c=a+(b+c)$	×	×	×	×
		Translative law: $\forall\ (a,b), \exists\ c$ such that $a+c=b$	×	×	×	×
	Rules of Product	Left distributive law: $a(b+c)=ab+ac$	×	×	×	×
		Right distributive law: $(a+b)c=ac+bc$	×	×	×	×
		Associative law: $(ab)c=a(bc)$	×	×	×	×
Unitary Ring		Identity element: $Ia=aI=a$	×	×	×	×
Division Ring		Non-zero elements have inverses: $a^{-1}a=aa^{-1}=I$	×	×	×	
Integral Domain		Commutative law: $ab=ba$	×	×		
Lineal Ring		Comparative law: one of ($a<b$, $a=b$, $a>b$) is true, the others are false.	×			
	Dimensionality:		1	2	4	variable

This set of division-ring number systems is unique; the first proof of the absence of any further finitely dimensioned division rings over the mathematical-real numbers is attributed to the work of Frobenius in 1878. However, our need for a division algebra to describe space does not necessarily extend beyond coordinate space to the quantitative description of individuated matter.

Clearly higher-tier generalizations don't recover specialized properties that were necessarily surrendered, by lower dimensional (lower tier) number systems, in order to expand the dimensionality of the algebra. Thus the argument for abandoning the commutation of products by quaternions holds for the next number-system generalization. Hence, we are led to conclude that (provided it is possible) we should make

$$\hat{e}_F^{(j_1)}\hat{e}_F^{(j_2)} + \hat{e}_F^{(j_2)}\hat{e}_F^{(j_1)} = 0 \qquad (10.31)$$

and

$$\hat{e}_F^{(j)}\hat{e}_{I'} + \hat{e}_{I'}\hat{e}_F^{(j)} = 0. \qquad (10.32)$$

We already have imposed the requirement $e_{I'}^2 = 0$. The anti-commutation is in conformity with this result and implies

$$\left(\hat{e}_F^{(j)}\right)^2 = 0. \qquad (10.33)$$

Thus we do not have to remove another further non-conformity. These conditions for generalizing the algebraic system, explain why the new designators form the basis of a graded Grassmann algebra rather than, for instance, an alternative Clifford algebra[9].

These Grassmann numbers have been introduced to enable a Fourier transformation of the aether. Thus Grassmann numbers have a physical significance which is quite different from the significance of the other numbers systems (listed in table 10.2) representing space. For the first time in the displayed hyper-complex sequence, we have a number system that does not include the previous number system as a specialized instance. Thus the Grassmann numbers must describe something other than space.

Furthermore, the designators are required to be of unit magnitude but (unlike quaternion and mathematical-complex numbers) their conjugates cannot be formed by flipping a mathematical sign (see equation (10.33)). Indeed the number system has also surrendered the requirement for an inverse. We can still construct a meaning for a unit magnitude by introducing the inverse Fourier transformation. This inverse Fourier transformation will introduce separate paired conjugate variables with which we can recover the concept of mathematical-real number lengths.

Now, an inverse Fourier transformation is very like a Fourier transformation. Thus the inverse Fourier transform will introduce a similar set of elements $(\hat{æ}_I, \hat{æ}_{I'}, \{\hat{æ}_F^{(j)}\})$ which form their own algebra and exhibit internally the same set of inter-relationships as the $(\hat{e}_I, \hat{e}_{I'}, \{\hat{e}_F^{(j)}\})$ namely

$$\hat{æ}_F^{(j_1)}\hat{æ}_F^{(j_2)} + \hat{æ}_F^{(j_2)}\hat{æ}_F^{(j_1)} = 0 \qquad (10.34)$$

$$\hat{æ}_F^{(j)}\hat{æ}_{I'} + \hat{æ}_{I'}\hat{æ}_F^{(j)} = 0 \qquad (10.35)$$

$$\hat{æ}_I \hat{æ}_F^{(j)} = \hat{æ}_F^{(j)} \hat{e}_I = \hat{æ}_F^{(j)} \qquad (10.36)$$

$$\hat{æ}_I \hat{æ}_{V'} = \hat{æ}_{V'} \hat{æ}_I = \hat{æ}_{V'} \qquad (10.37)$$

$$\hat{æ}_I^2 = \hat{æ}_I \,. \qquad (10.38)$$

However, the two algebrae need to be able to work together to make an inverse Fourier transform reverse the Fourier transform with no loss of information. If there is no loss of information, then the Fourier transformed and the untransformed representations must be equally valid representations of reality.

Traditionally when one number system encompasses another smaller number system, we may call upon the procedure known as 'analytical continuation' to help generalize a useful function in the smaller number system to obtain a corresponding function in the larger number system. For example, the mathematical-real numbers encompass the integers. The exponential function is a well-defined function for integers. Here exponentiation corresponds to repeated multiplication and perhaps an inverse multiplication (ie division). An analytical expression for the exponential function of a mathematical-real number having an infinitesimal fractional part is directly obtainable from known functional relationships for integer powers. Presuming that a unique mathematical-real analytical expression exists for an exponential function, the deduced analytical form can be analytically extended to all real numbers. However, general exponentiation is not unique (it is a many-one function) even for non-integer rational powers. For example, the raising an integer to power one-half has two potential solutions; raising an integer to an irrational power has an infinite number of solutions. For the most part, we overcome this multiplicity by choosing a unique principal-value function.

We need to take even more care with Grassmann numbers since they do not encompass the quaternion numbers. Hence the Grassmann zero is not common to a 'lower' number scheme. The Grassmann dimensions are associated with the analytical requirements of taking the quasi-scalar products. Thus as the magnitude of the quasi-number asymptotes to zero, we do not lose the dimension. Hence, in place of an identity for an exponential of (a Grassmann) zero we find something else

$$\underset{a \to 0}{Lt} \exp(a\hat{i}\hat{\gamma}_0 \hat{e}_F^{(j)}) = (\hat{r}\hat{\gamma}_0)^{\hat{e}_F^{(j)}} = \hat{E}_F^{(j)} \qquad (10.39)$$

$$\exp(a\hat{i}\hat{\gamma}_0 \hat{æ}_F^{(j)}) = (\hat{r}\hat{\gamma}_0)^{\hat{æ}_F^{(j)}} = \hat{Æ}_F^{(j)} \text{ for } a = 0\,. \qquad (10.40)$$

Note that the $\hat{E}_F^{(j)}$ share the identity designator $\hat{e}_I$; the $\hat{Æ}_F^{(j)}$ share the identity $\hat{æ}_I$. A power expansion of the exponential of a quasi-scalar product must terminate at first order. Hence, we shall define the exponential function by the power series

$$\exp(a\hat{i}\hat{\gamma}_0\hat{e}_F^{(j)}) = \hat{E}_F^{(j)} + a\hat{i}\hat{\gamma}_0\hat{e}_F^{(j)} \qquad (10.41)$$

$$\exp(a\hat{i}\hat{\gamma}_0\hat{æ}_F^{(j)}) = \hat{Æ}_F^{(j)} + a\hat{i}\hat{\gamma}_0\hat{æ}_F^{(j)} \qquad (10.42)$$

Then the inverse cancellation of exponentials for the same particle

$$æxp(-\hat{i}\hat{\gamma}_0\hat{æ}_F^{(j)}a)\exp(a\hat{i}\hat{\gamma}_0\hat{e}_F^{(j)}) = (\hat{Æ}_F^{(j)} - a\hat{i}\hat{\gamma}_0\hat{æ}_F^{(j)})(\hat{E}_F^{(j)} + a\hat{i}\hat{\gamma}_0\hat{e}_F^{(j)}) \qquad (10.43)$$

is accomplished if $\hat{E}_F^{(j)}$ and $\hat{Æ}_F^{(j)}$ satisfy

$$\hat{Æ}_F^{(j)}\hat{E}_F^{(j)} = \hat{e}_I \,. \qquad (10.44)$$

The first transformation completed cannot be treated as an inverse transformation. An inverse transformation is only meaningful after a transformation has been made. Thus we may write

$$\hat{E}_F^{(i)}\hat{Æ}_F^{(j)} = 0 \qquad \text{and even } \hat{Æ}_F^{(i)}\hat{E}_F^{(j)} = \delta_{i,j}\hat{e}_I \,. \qquad (10.45)$$

These mixed-system results apply also to the identities

$$\hat{æ}_I\hat{e}_I = \hat{e}_I \qquad \hat{e}_I\hat{æ}_I = 0 \,. \qquad (10.46)$$

The $\hat{E}_F^{(j)}$ appear to be an alternative (capital) representation of the set of designators $\hat{e}_F^{(j)}$. To the extent that we can generate either one of these sets from the other set; if the $\hat{e}_F^{(j)}$ anti-commute, then the $\hat{E}_F^{(j)}$ cannot commute (and likewise for the conjugate system). Indeed, we must require $\hat{E}_F^{(j)}$ to anti-commute

$$\hat{E}_F^{(i)}\hat{E}_F^{(j)} + \hat{E}_F^{(j)}\hat{E}_F^{(i)} = 0 \qquad i \neq j \qquad (10.47)$$

$$\hat{Æ}_F^{(i)}\hat{Æ}_F^{(j)} + \hat{Æ}_F^{(j)}\hat{Æ}_F^{(i)} = 0 \quad i \neq j \,. \qquad (10.48)$$

Next, we recognize that diphthong designators are like destruction operators and non-ligated designators are like creation operators. Hence, we shall require a form of commutation for the (double-creation) mixed algebra.

$$\hat{æ}_F^{(j)}\hat{E}_F^{(i)} = Æ_F^{(j)}\hat{e}_F^{(i)} \qquad (10.49)$$

However, the normalization condition actually requires that the capitalized and non-capitalized designators to be the same. These results will deliver the anticipated loss-less transformation. Consider the expression

$$\exp\left(-\hat{i}\,\bar{\bar{q}}\underline{\underline{\xi}}^{(i)}\right)\exp\left(\hat{i}\,\bar{\bar{q}}\underline{\underline{\xi}}^{(j)}\right) = \hat{r}\hat{\gamma}_o(æ_F^{(i)} - \hat{i}\,\bar{\bar{q}}\underline{\underline{\xi}}^{(i)})(e_F^{(j)} + \hat{i}\,\bar{\bar{q}}\underline{\underline{\xi}}^{(j)}) \qquad (10.50)$$

appearing in an inverse transformation recovery equation. Then we can expand the brackets obtaining terms of Grassmann grade 0, 1 and 2.

$$\exp\left(-\hat{i}q\underline{\underline{\xi}}^{(i)}\right)\exp\left(\hat{i}q\underline{\underline{\xi}}^{(j)}\right)=\hat{r}\hat{\gamma}_o\left(æ_F^{(i)}e_F^{(j)}+q\underline{\underline{\xi}}^{(i)}q\underline{\underline{\xi}}^{(j)}\right) +\hat{i}\left(æ_F^{(i)}q\underline{\underline{\xi}}^{(j)}+(-)_T e_F^{(j)}q\underline{\underline{\xi}}^{(i)}\right) \quad (10.51)$$

Hence it follows that

$$\exp\left(-\hat{i}q\underline{\underline{\xi}}^{(i)}\right)\exp\left(\hat{i}q\underline{\underline{\xi}}^{(j)}\right)=\hat{r}\hat{\gamma}_o\left(\hat{e}_I\delta_{i,j}+q\underline{\underline{\xi}}^{(i)}q\underline{\underline{\xi}}^{(j)}\right) +\hat{i}q\left(\underline{\underline{\xi}}^{(j)}+(-)_T\underline{\underline{\xi}}^{(i)}\right)\delta_{i,j}\hat{e}_I. \quad (10.52)$$

When $i=j$, the product of exponentials gives a consistent coherent magnitude of unity. Otherwise, we are unlikely to see much coherence between the various terms. Thence we deduce that the average of this product, when taken over a large reciprocal latticework of vectors, yields a Kronecker delta function

$$\frac{1}{N}\sum_{q}\exp\left(-\hat{i}q\underline{\underline{\xi}}^{(i)}\right)\exp\left(\hat{i}q\underline{\underline{\xi}}^{(j)}\right)=\hat{r}\hat{\gamma}_0\hat{e}_I\delta_{i,j}+O(N^{-½})\cdot \quad (10.53)$$

Note that anti-commutation implies that the exponentials anticommute for $i\neq j$.

$$\exp\left(-\hat{i}q\underline{\underline{\xi}}^{(i)}\right)\exp\left(\hat{i}q\underline{\underline{\xi}}^{(j)}\right)+\exp\left(\hat{i}q\underline{\underline{\xi}}^{(j)}\right)\exp\left(-\hat{i}q\underline{\underline{\xi}}^{(i)}\right)=2\hat{r}\hat{\gamma}_0\hat{e}_I\delta_{i,j} \quad (10.54)$$

Equation (10.53) is the critical equation; this equation shows that quasi-scalar products can discharge their duty by delivering a lossless reciprocal-space transformation that leads to an algebraic system of collective coordinates.

10.3 Collective Transformations

Now, we may proceed by creating the collective transformation components of the macroscopically averaged aethereal field. The transformations are performed in the mathematical-complex SU(2) subspace. Thus we divide complex 3-space (fixed time, t, and fixed specific energy, w_0) into a large number of equal-size cells as determined by a chosen processing scale. This procedure is performed in identical ways on each branch ie we have the same distribution of cells on each branch.

For each element of the matrix, $\underline{\underline{\Phi}}_I^{m_u,n_u}$, we construct a set of collective field elements of a form

$$a_{\underline{\underline{q}}}(w_0,t)=\sum_j \Phi_{\underline{\underline{l}}}^{m_u, n_u}(\underline{\underline{\xi}}^{(j)}) \exp\left(\hat{i}\,\underline{\underline{q}}\,\underline{\underline{\xi}}^{(j)}\right). \qquad (10.55)$$

Here the index of summation, j, runs over all coordinate cells on branch l. Note that this summation could be indexed separately in the six different coordinate directions. Furthermore each reciprocal vector, $\bar{\bar{q}}$, in our reciprocal transformation is a quasi-number.

The collective coordinates, $a_{\underline{\underline{q}}}$, are fields spanning quasi-space. For notational simplicity, in the following equations, often we shall not display the dependence of the collective coordinate upon the indices m_u, n_u and l. The summation with index j is taken over cells in a finite isolated region of a branchless mathematical-complex quaternion space. Implicitly the origin of relative coordinates is to be located near the centre of this region and periodic boundary conditions are assumed to apply at the boundary of this region. If this region were a hyper-cube, then it would be appropriate to use Cartesian coordinates. For instance, if the x-position coordinate-component, ξ_x, in mathematical-complex quaternion space runs from $-HL_x/2c$ to $HL_x/2c$ as

$$\xi_x=\frac{H}{c}\times\frac{\kappa L_x}{N_x} \quad \text{for } \kappa=0,\pm 1,\pm 2\ldots\pm N_x/2, \qquad (10.56)$$

the corresponding reciprocal coordinate would take values

$$q_{W_x}=\frac{1}{m_0 c}\times\frac{h\kappa}{L_x} \quad \text{for } \kappa=0,\pm 1,\pm 2\ldots\pm N_x/2. \qquad (10.57)$$

Here h is a constant, which we shall show to be a part of the foundation of our measurement-unit system. It is relevant to note that this reciprocal vector exists in a quasi-space that is quite similar in form to the (vector-dimensioned) foundation space spanning the mathematical-complex quaternions. Effectively in the Fourier-style transformation those vector components, that are mathematical-real in one space, have been interchanged with the vector components that are mathematical-imaginary in the other space. The physics that connects such identifications will be discussed in the next chapter.

Now then, suppose that we express our mathematical-complex quaternion spatial variable (equation (9.3)) as

$$\underline{\underline{\xi}}^{(j)}=\hat{\underline{\gamma}}.\left(\hat{r}\,\underline{w}+\hat{i}(H/c)\underline{x}\right)+\gamma_0\left(\hat{r}w_0+\hat{i}Ht\right) \qquad (10.58)$$

($\hat{\underline{\gamma}}$ was defined in equation 10.17) and that we write the quasi-space vector as

$$\overline{\overline{q}} = \underline{\hat{\gamma}}.\left(\hat{r}(H/c)\overline{X} + \hat{i}\overline{W}\right). \qquad (10.59)$$

Here $\overline{X}$ is to be the Fourier-conjugate variable of $\underline{w}$, and $\overline{W}$ is to be the Fourier-conjugate variable of $\underline{x}$. Then the required complex-real scalar product, the quasi-scalar product of equation (10.5), is

$$\begin{aligned}&\left(\hat{r}\underline{w} + \hat{i}(H/c)\underline{x} + \hat{r}w_0 + \hat{i}Ht\right)\left(\hat{r}(H/c)\overline{X} + \hat{i}\overline{W}\right)\\ &= (-)_P \hat{r}\hat{\gamma}_0(H/c)\left(\underline{w}\overline{X} + (-)_T \underline{x}\overline{W}\right)\hat{e}_F^{(j)}.\end{aligned} \qquad (10.60)$$

By design, there are no complex-imaginary cross-terms in this complex-real *scalar* product and the time and specific energy dependencies are not transformed. In this product, the complex-imaginary symbol, $\hat{i}$, is part of a vector-component label, ie it is a designator factor in an eight-dimensional mathematical-real space.

By virtue of equation (10.53), we can perform a reverse Fourier transformation. Thus for a large number of cells $N = N_x N_y N_z$, this reverse transformation should give an excellent approximation to the continuum aethereal field in terms of the collective coordinates,

$$\underline{\underline{\Phi}}_l^{m_u n_u}(\underline{\underline{\xi}}) = \frac{1}{N}\sum_q \underline{\underline{a}}_{\overline{\overline{q}}} \exp\left(-\hat{i}\,\overline{\overline{q}}\,\underline{\underline{\xi}}\right). \qquad (10.61)$$

10.4 Local Reality

Experience informs us that we rely upon a local reality to make sense of the world. However, locality is important with or without our contingent existence. Recall that all aethereal qualities are derived from the local relationships between events. Hence, if all the elements of a specific reality were represented in fully delocalized terms, then there would be very little capacity to distinguish natural features within the reality. This conclusion does not imply that reality must be either partially or totally local. The *validity* of every representation of eventual reality is on an equal footing with every other potential way to describe reality. However, the conclusion implies that the content of a partially localized representation of reality can be rich and interesting, whereas, the content of a delocalized representation will be impoverished and dull. The optimum representation for us to use is that one, which will tell us most about reality, ie that one which generates the richest interpretations. This is the rationale for seeking a substantial measure of locality in the representations of theory of eventual reality. Consequently hierarchical descriptions of reality should not be treated as an unchangeable aspect of the universe; after all, reality is what natural philosophy and science says reality is – reality is partly subjective in that accepted reality is era dependent.

A likely local representation of reality imparts several expectations; these are expectations that there should be assignable localized elements of reality that interact locally. However, there are many participatory concepts that are inherent in a description of locality. Some aspects of locality will be introduced; other aspects of locality will emerge naturally. Hence we will find that locality is central in constructing the analytical elements of physical science and the hierarchy of scientific worldviews.

One expectation (local isolation) is that it is meaningful to study isolated subsystems as though they were an entire system. This assumption has already been made. Another expectation (local constituency) is that we may continue to break the description of reality down into smaller and smaller component elements that are increasingly more fundamental until the elements become indivisible. This idea is compatible with the reductionist view of our edifice of reality with its hierarchy of storeys leading down to an ultimate foundation. Another expectation (local legislation) is that we may associate physical attributes and rules to all isolated subsystems. Another expectation (local cumulation) is that there are system quantities (like classical momenta or energies) for which the sum of individual contributions within the subsystem agree with value of this quantity for the subsystem expressed without explicit knowledge of the internals of the subsystem. Another expectation (local influence) is that, as all explicit influences are to be local, high-order derivatives (beyond some finite order) will not feature in explicit equations for physical quantities. Each of these expectations has been given a brief reference phrase as a useful label to help make clear which aspects of locality are being considered in the following arguments.

In the hierarchy of scientific representations (the storeys of the edifice of reality), quantum mechanics sits between the coordinate-space representation of eventual reality and classical mechanics. Amongst its several distinguishing features, classical mechanics is inherently local. Quantum mechanics is partially local, since quantum mechanics does import some features of classical reality. We have already developed (in the previous sections) the mathematical contortions required to forge this partially local picture of reality from an inherently a-local basis. Next, we show that these contortions naturally lead to some of the familiar representations of quantum mechanics.

We have constructed the collective-field coordinates as

$$\underline{\underline{a}}_{\bar{\bar{q}}} = \sum_{j} \underline{\underline{\Phi}}_{l}^{m_u, n_u} \exp\left(\hat{i}\, \bar{\bar{q}}\, \underline{\underline{\xi}}^{(j)} \right) \qquad (10.62)$$

in a quasi-space (or reciprocal Fourier space). The power-series expansion of the exponential has only two terms:

$$\underline{\underline{\Phi}}_{l}^{m_u, n_u}(\underline{\underline{\xi}}^{(j)})\left(\hat{E}_F^{(j)} + \hat{i}\, \bar{\bar{q}}\, \underline{\underline{\xi}}^{(j)} \right) = \underline{\underline{\Phi}}_{l}^{m_u, n_u}(\underline{\underline{\xi}}^{(j)})\hat{E}_F^{(j)} + \underline{\underline{\phi}}^{(j)}(\bar{\bar{q}}; w_0, t)\hat{e}_F^{(j)}. \qquad (10.63)$$

These two terms represent the separation of a zero-point field in empty space and an aethereal structure in space that is bound up with the formalism of quantum mechanics. The spatial quasi-number dependence is found only in the second reified-matter term. The first term, which could be said to contribute equally to all 'locations' in quasi-space, is a description of empty quasi-space ie without the quasi-spatially varying reified-matter fields, $\underline{\underline{\phi}}^{(j)}(\overline{\overline{q}};w_0,t)$. The quasi-spatially varying fields are given by

$$\underline{\underline{\phi}}^{(j)}(\overline{\overline{q}};w_0,t)\hat{e}_F^{(j)} = (-)_P \underline{\underline{\Phi}}_l^{m_u n_u}(\underline{\underline{\xi}}^{(j)})\left(\sum_{i=1..3} q_i\xi_i^{(j)} + (-)_T \sum_{i=5..7} q_i\xi_i^{(j)}\right)\hat{\gamma}_0\hat{i}\,. \quad (10.64)$$

Now, quasi-space has only six dimensions; however, it is convenient to expand quasi-space to the full eight dimensions by adding new time and specific-energy dimensions. This 8-space will be called hybrid space. However, we also recognize that the quasi-space dimensions have particular designators $\{\hat{e}_F^{(j)}\}$ associated with each particle field. Hence, we create hybrid space with eight available dimensions for each particle by appending $\hat{e}_V$ to the time and specific energy dimensions but retaining the magnitudes of coordinate space

$$T = t \text{ and } W_0 = w_0\,. \quad (10.65)$$

Clearly these new dimensions are not reciprocally transformed (ie time is transformed to time and specific energy is transformed to specific energy).

Table 10.3

Notational conventions that apply to the unit-dimensionless coordinates of a single-particle hybrid space.

	Designator	Hybrid Space
0	$\hat{r}\hat{\gamma}_0\hat{e}_V$	q_0, W_0 or E/m_0c^2
1	$\hat{r}\hat{\gamma}_1\hat{e}_F^{(j)}$	q_1, HX_1/c
2	$\hat{r}\hat{\gamma}_2\hat{e}_F^{(j)}$	q_2, HX_2/c
3	$\hat{r}\hat{\gamma}_3\hat{e}_F^{(j)}$	q_3, HX_3/c
4	$\hat{i}\hat{\gamma}_0\hat{e}_V$	q_4 or HT
5	$\hat{i}\hat{\gamma}_1\hat{e}_F^{(j)}$	q_5, W_1 or P_1/m_0c
6	$\hat{i}\hat{\gamma}_2\hat{e}_F^{(j)}$	q_6, W_2 or P_2/m_0c
7	$\hat{i}\hat{\gamma}_3\hat{e}_F^{(j)}$	q_7, W_3 or P_3/m_0c

The factor $\hat{e}_V$ (as defined by equation (10.16)) signifies that the quaternion-scalar dimensions relate to products of the other quasi-numbers. The inclusion of this designator brings all the dimensions of hybrid space into a common number system described by the set of new designators. The expanded eight-dimensional quasi-space (for each particle) is referred to as hybrid space. Variables within hybrid space are denoted by the symbol $\tilde{q}$. The new untransformed dimensions of hybrid space (T and W_0) complement and mirror the algebraic relations of the old transformed hybrid-space dimensions. For instance, the products of four hybrid-space variables, $\tilde{q}$, are always zero. Thus using hybrid-space notation, equation (10.63) would be written as

$$\underline{\underline{\Phi}}_I^{m_u,n_u}(\underline{\underline{\xi}}^{(j)})\left(\hat{E}_F^{(j)}+\hat{i}\,\overline{\overline{q}}\,\underline{\underline{\xi}}^{(j)}\right)=\underline{\underline{\Phi}}_I^{m_u,n_u}(\underline{\underline{\xi}}^{(j)})\hat{E}_F^{(j)}+\underline{\underline{\phi}}^{(j)}(\tilde{q})\hat{e}_F^{(j)}. \qquad (10.66)$$

The fields, $\underline{\underline{\phi}}^{(j)}(\tilde{q})$, provide a basis for the local-constituency quantum worldview. This set of fields in quasi-space holds the spatially varying details of the underlying reality of the aethereal field. Members of this set of fields will be called particle fields. (The purists may prefer to restrict such terminology to products of these fields on sets of same-time-sense branches).

The construction of cumulative physical properties for a local reality in quasi-space also relies upon a restriction of the algebra satisfied by the particle-field variables. If candidate physical properties have more than a linear dependence on these particle-coordinate variables, then system-constructed properties would involve cross terms of the particle coordinates such as $\underline{\underline{\phi}}^{(j_1)}\hat{e}_F^{(j_1)}\,\underline{\underline{\phi}}^{(j_2)}\hat{e}_F^{(j_2)}$ (for $j_1 \neq j_2$). For all expressions giving equal weighting to each particle field, the cross-terms cancel generating the impression that the whole is the sum of the properties of its particles.

This symmetric characteristic satisfies the requirements for local cumulation. Hence, it is seen that the imposition of quasi-number transformations allows us to form particle fields that can provide a partially local description of reality.

10.5 Mechanics

Now, we shall see how we can predict the time evolution of the particle fields in hybrid space. Hence we will identify hybrid space with the space of our scientific awareness. We start the analysis by considering an entropy parameter that describes the order or disorder in the aethereal field of hybrid space. Within our supposedly isolated subsystem of the universe, this entropy parameter is described by some functional of the aethereal field. However, the variations of the aethereal field will be represented in local-constituency form by sets of particle fields in hybrid space. Thus we express the hybrid-space entropy parameter as an unknown functional of the set of particle fields

$$\underline{\underline{S}} = \underline{\underline{S}}\left[\{\underline{\underline{\phi}}^{(j)}(\tilde{q})\}\right] . \tag{10.67}$$

Here the square brackets signify a functional relationship (in other equations square brackets accompanied by a subscripted sign is used to signify a commutator or anti-commutator) and the curly brackets are used to signify that the functional has a dependence upon the *set* of particle fields. We will suppose that we may investigate one ground-level branch at a time. Consequently we will consider one set of values for l, n and m.

Now, a general non-local functional can be written as a spatial integral. For our example, the integration will be performed in hybrid space. In general, the integrand is expandable as a series of terms that contain combinations of the particle function and spatial derivatives of the particle function; these terms are indexed by the highest order of hybrid-spatial derivative appearing within the term.

Now, we shall denote the first derivatives of the particle fields (with respect to spatial component *magnitudes*) as

$$\underline{\underline{\phi}}^{(j)}_{,D\lambda}(\tilde{q}) = \frac{\partial \underline{\underline{\phi}}^{(j)}(\tilde{q})}{\partial q_\lambda} . \tag{10.68}$$

Here the absence of bars and tildes on the variable of differentiation indicates that q_λ is an argument magnitude ie q_λ is free of any designators.

In our non-tensorial notation, we introduce the capital D label as shorthand notation to indicate differentiation. This additional label avoids confusion with mathematical-complex quaternion component indices. Thus, suppressing branch indices, we may also write

$$\phi^{(j)}_{\upsilon,D\lambda}(\tilde{q}) = \frac{\partial \phi^{(j)}_{\upsilon}(\tilde{q})}{\partial q_\lambda} . \tag{10.69}$$

Let us firstly consider the six quasi-space components of this derivative field. The differentiation in quasi-space is straightforward and we obtain a quasi-space vector with numerical value

$$\underline{\underline{\phi}}^{(j)}_{,D\lambda}(\tilde{q}) = (-)_P(\pm)_T \hat{i}\,\xi^{(j)}_\lambda \underline{\underline{\Phi}}^{m_u,n_u}_{l}(\underline{\underline{\xi}}^{(j)}) \exp\left(\hat{i}\,\overline{\overline{q}}\,\underline{\underline{\xi}}^{(j)}\right) \hat{e}^{(j)}_F$$
$$\text{for } \lambda \neq 0, \lambda \neq 4 . \tag{10.70}$$

Here the $(\pm)_T$ is negative for λ less than 4 and positive otherwise. Furthermore, the second-order derivatives of a particle-field are zero. Note that each derivative in a *quasi-direction* provides a factor, $\hat{e}^{(j)}_F$; thus second derivatives in quasi-directions must vanish.

$$\frac{\partial^2 \underline{\underline{\phi}}^{(j)}(\tilde{q})}{\partial q_{\lambda_1} \partial q_{\lambda_2}} = (-)_T \xi_{\lambda_1}^{(j)} \xi_{\lambda_2}^{(j)} \underline{\underline{\Phi}}_I^{m_u, n_u}(\underline{\underline{\xi}}^{(j)}) \exp\left(\hat{i}\bar{\bar{q}}\underline{\underline{\xi}}^{(j)}\right) \hat{e}_F^{(j)2} \qquad (10.71)$$

However, the inclusion, of the $\hat{e}_{I'}$ designator in the definition of the dimensions representing hybrid time and hybrid specific-energy, also introduces a factor of $\hat{e}_{I'}$ with each differentiation in these scalar directions. Hence these designators preserve the orthogonal spatial symmetry of hybrid space (equation (10.32)). With this symmetry, the full second-order derivatives of $\underline{\underline{\phi}}^{(j)}(\tilde{q})$ are seen to vanish.

Likewise, we observe that the square of a first derivative, $\underline{\underline{\phi}}^{(j)}_{,D\lambda}(\tilde{q})$, always vanishes. Furthermore, symmetric products of first derivatives also cancel

$$\underline{\underline{\phi}}^{(j_1)}_{,D\lambda_1}(\tilde{q})\underline{\underline{\phi}}^{(j_2)}_{,D\lambda_2}(\tilde{q}) + \underline{\underline{\phi}}^{(j_2)}_{,D\lambda_2}(\tilde{q})\underline{\underline{\phi}}^{(j_1)}_{,D\lambda_1}(\tilde{q}) = 0 . \qquad (10.72)$$

These mathematical rules inform us as to how any functional dependency can be expressed: there are no second derivatives, there are only first derivatives raised to first-order powers; there are no restrictions on the powers of non-differentiated particle-field terms.

We seek a local-constituency formulation for the systems of particles fields. Local constituency implies a spatially separable reality, a reality in which complex systems are just an arrangement and superposition of simpler entities. However, supposition of elemental solutions implies the appropriate fundamental equations are linear equations. A linear formulation is accomplished by iteratively introducing new fields. Thus to accommodate the allowed higher powers of the particle fields and to explicitly segregate potentially commuting behaviour in the functional, in the first instance, we shall introduce a new set of fields, $\underline{\underline{B}}^{(j)}$, which are related to the square of a particle field by the equation

$$\phi_{\lambda_1 \circ \lambda_2}^{m_1 \circ m_2, n_1 \circ n_2, (j)} B_{\lambda_2}^{m_2, n_2, (j)} \phi_{\lambda_1}^{m_1, n_1, (j)} = \hat{r}\hat{\gamma}_0 \hat{e}_I . \qquad (10.73)$$

Here quaternion rules of multiplication are used to select the appropriate component elements that are involved. Thus the product, which we have indicated by operation $\circ$, is specified according to the rules given in table 10.4.

Physically the process of separating out field terms involving higher powers corresponds to the process of 'undressing' the particle field to reveal a single fermion that is free of a cloud of virtual bosons. These new fields, $\underline{\underline{B}}^{(j)}$, will be referred to as dressed-boson fields. Later this 'undressing' will be repeated iteratively to identify individual bosons. Thus it is our quest for superposable states that reveal the character of identical-particle statistics.

Table 10.4

Rule for assigning index, $n_1 \circ n_2$, obtained when multiplying two mathematical-complex quaternions with indices n_1 and n_2.

n_1 \ n_2	0	1	2	3	4	5	6	7
0	0	1	2	3	4	5	6	7
1	1	0	3	2	5	4	7	6
2	2	3	0	1	6	7	4	5
3	3	2	1	0	7	6	5	4
4	4	5	6	7	0	1	2	3
5	5	4	7	6	1	0	3	2
6	6	7	4	5	2	3	0	1
7	7	6	5	4	3	2	1	0

It may seem strange to define $\underline{\underline{B}}^{(j)}$ fields in terms of an *inverse* of the product of particle fields, $\underline{\underline{\phi}}^{(j)}(\tilde{q})$; nevertheless it will be shown that this choice delivers the more symmetric formalism. As with the original particle fields, $\underline{\underline{\phi}}^{(j)}(\tilde{q})$, the second derivative of these new fields and the square of the first derivative both vanish. For those equations in which the higher powers are absorbed into the dressed-boson fields, the $\underline{\underline{\phi}}^{(j)}(\tilde{q})$ fields will be referred to as fermion-particle fields.

Consequently *only* the fermion and dressed-boson fields and their first derivatives can feature in a functional expansion in hybrid space. Furthermore, the functional expansion will only involve linear terms in the fermion particle fields (higher order powers are incorporated in the dressed-boson particle fields) and linear terms in fermion individual-particle first derivatives.

Explicitly we write

$$S_{k\,j}^{m,n} = \int d\tilde{q}\ \Lambda_{k,l}\left(\{\underline{\underline{\phi}}^{(j)}\},\{\underline{\underline{\phi}}^{(j)}_{,D\lambda}\},\{\underline{\underline{\mathrm{B}}}^{m_1,n_1,(j)}\},\{\underline{\underline{\mathrm{B}}}^{m_2,n_2,(j)}_{,D\lambda}\}\right). \qquad (10.74)$$

The distinctive form of this equation is known from classical mechanics. This distinctive form is a familiar starting point for Lagrangian field theory. In this work, this form has been engineered by the requirements of the quasi-number algebra. Next, this equation will be used to determine a single-particle equation-of-motion for a fermion. To a significant extent, the form of this equation does rely upon the initiating algebra of eventual reality.

10.6 Variational Principle

Now, we reapply Fermat's principle of stationariness which, in this context, is commonly referred to as the variational principle. We may assert that the entropy parameter (which varies monotonically with entropy) has an extreme value; however, that would be to presume the second law of thermodynamics. Thus, more appropriately, we say that the dominant/realizable particle fields are those that coherently add their contributions to the entropy parameter. Either way, the integral is required to be stationary for any infinitesimal localized variation in a particle field. In part 3 we shall extend further this stationariness to space.

Let us consider a fluctuation in a fermion field; this should leave the entropy parameter unchanged. Technically the fermion field and the dressed-boson field are related by equation (10.73). However, the method of Lagrange multipliers enables us to freely vary the fermion fields (keeping the dressed-boson field a constant) and account for the interdependency as a system constraint.

Thus we mirror the Lagrangian derivation of classical field theory and performing a functional variation on a single fermion particle field. For a fluctuation in the fermion field of $\delta\phi_{\upsilon}^{(j)}$ we have

$$\delta \underline{\underline{S}} = \int d\tilde{q} \left(\frac{\partial \underline{\underline{\Lambda}}}{\partial \phi_{\upsilon}^{(j)}} \delta\phi_{\upsilon}^{m_1,n_1(j)} + \sum_{\lambda} \frac{\partial \underline{\underline{\Lambda}}}{\partial \phi_{\upsilon,D\lambda}^{(j)}} \delta\phi_{\upsilon,D\lambda}^{m_1,n_1(j)} \right. \\ \left. + \sum_{\lambda, m_2, n_2} g_1^{m_1 \circ m, n_1 \circ n} \phi_{\lambda\circ\upsilon}^{m_2,n_2,(j)} B_{\lambda}^{m_2,n_2(j)} \delta\phi_{\upsilon}^{m_1,n_1(j)} \right) \qquad (10.75)$$

where $g_1^{m,n}$ are Lagrange multipliers. Here the index, λ, runs over all eight of the hybrid dimensions. We are attempting to derive a simple one-particle equation of motion to resemble the Dirac equation. Henceforth for simplicity, we shall ignored couplings that change a particle's quantum numbers. Then the boson field will become a photon field. Of course all couplings should be retained in a many-body formulation.

Now, we perform a partial integration and note that

$$\delta\underline{\underline{\phi}}_{,D\lambda}^{(j)} = \frac{\partial}{\partial \tilde{q}_{\lambda}} \delta\underline{\underline{\phi}}^{(j)}(\tilde{q}) \cdot \qquad (10.76)$$

Every fluctuation must leave the entropy parameter unchanged. Hence, we obtain an Euler's equation

$$\frac{\partial \Lambda_{kJ}}{\partial \phi_{\upsilon}^{(j)}} - \sum_{\lambda} \frac{\partial}{\partial \tilde{q}_{\lambda}} \left(\frac{\partial \Lambda_{kJ}}{\partial \phi_{\upsilon,D\lambda}^{(j)}} \right) + \sum_{\lambda} 2 g_1 \phi_{\lambda}^{(j)} B_{\lambda\circ\upsilon}^{(j)} = 0 \qquad (10.77)$$

where we suppress the pair of zero superscripts on g_1 and $B_\upsilon^{(j)}$. On the left-hand side of this equation, the first two terms are the regular variational terms and the last term accounts for the algebraic connection (ie interaction) between fermion and dressed-boson fields.

We solve equation (10.77) by firstly looking for a solution of the complementary equation

$$\frac{\partial\Lambda_{kJ}}{\partial\phi_\upsilon^{(j)}}-\sum_\lambda\frac{\partial}{\partial\tilde{q}_\lambda}\left(\frac{\partial\Lambda_{kJ}}{\partial\phi_{\upsilon,D\lambda}^{(j)}}\right)=0\cdot \qquad (10.78)$$

The complementary equation has solutions made from products of powers of $\phi_\upsilon^{(j)}(\tilde{q})$ and $\phi_{\upsilon,D\lambda}^{(j)}(\tilde{q})$

$$\begin{gathered}\left(\frac{\partial}{\partial\phi_\upsilon^{(j)}}-\sum_{\lambda'}\frac{\partial}{\partial\tilde{q}_{\lambda'}}\left(\frac{\partial}{\partial\varphi_{\upsilon,D\lambda'}^{(j)}}\right)\right)\left(\phi_\upsilon^{(j)}(\tilde{q})\right)^s\left(\phi_{\upsilon,D\lambda}^{(j)}(\tilde{q})\right)^r\\ =(s-rs)\left(\phi_\upsilon^{(j)}(\tilde{q})\right)^{s-1}\left(\phi_{\upsilon,D\lambda}^{(j_2)}(\tilde{q})\right)^r.\end{gathered} \qquad (10.79)$$

Thus the complementary solution may be made from terms with either $s=0$ or $r=1$. Furthermore, the expression $g_1\phi_{\lambda\circ\upsilon}^{(j)}B_\lambda^{(j)}\phi_\upsilon^{(j)}$ will serve as a particular solution. Hence, we note that the Euler's equation for fermions holds when Λ takes the general form

$$\Lambda_\upsilon=\hat{d}_0+\phi_\upsilon^{(j)}\left\{\sum_\lambda\hat{c}_\lambda\phi_{\upsilon,D\lambda}^{(j)}-g_1B_{\lambda\circ\upsilon}^{(j)}\phi_\lambda^{(j)}\right\}+\sum_\lambda\hat{f}_{\upsilon,\lambda}\phi_{\upsilon,D\lambda}^{(j)} \qquad (10.80)$$

where we have introduced constants, $\hat{d}_0$, $\hat{c}_\lambda$ and $\hat{f}_{\upsilon,\lambda}$. Now, the $\hat{d}_0$ term is an unimportant constant that sets the zero level that defines the entropy of matter-free space. We shall follow common practice and require that $\underline{\underline{\Lambda}}$ vanish on the boundary of the isolated region. Therefore we set $\hat{d}_0=0$. Actually a near zero value (of $\underline{\underline{\Lambda}}$) across all space might be expected when an almost infinite range of choices is available.

Furthermore the entropy parameter should not be dependent upon the sign convention associated with particle fields ie flipping the sign of all fermion fields should leave the entropy parameter unchanged. Thus we exclude the odd-power $\hat{f}_{k,l\lambda}$ terms.

Consequently we deduce that Λ has the general form

$$\Lambda_\upsilon=\phi_\upsilon^{(j)}\left\{\sum_\lambda\hat{c}_\lambda\phi_{\upsilon,D\lambda}^{(j)}-g_1B_{\lambda\circ\upsilon}^{(j)}\phi_\lambda^{(j)}\right\}\cdot \qquad (10.81)$$

Unlike the classical derivation of quantum mechanics, substituting this form into Euler's equation does not yield an equation of motion. Instead we will impose a condition that Λ should vanish everywhere. The reasoning is simple: quantum mechanics is a regional theory; it is not applicable to the entire universe. Consistency of interpretation requires that every region be able to be divided into smaller subregions without disturbing the interpreted reality. Thus we require that $\underline{\underline{\Lambda}}$ be constant everywhere. Physically this condition corresponds to assuming that entropy is maximal with or without the matter fields which we may freely superimpose to create instances of local reality. Hence we place everywhere in hybrid space on an equal footing. $\underline{\underline{\Lambda}}$ does not only vanish on the boundaries of the 'isolated' region. However, the most satisfactory explanation (derived from a variation in the scale of space) of this requirement is left to equation (17.12).

This maximal condition is achieved when the individual fermion particle fields satisfy a first order differential equation of the form given by

$$\sum_{\lambda} \hat{c}_{\lambda}\phi^{(j)}_{\upsilon,D\lambda} = g_1 B^{(j)}_{\upsilon\circ\lambda}\phi^{(j)}_{\lambda}. \qquad (10.82)$$

This result describes the form of an equation-of-motion in hybrid space for our fermion particle, j, on branch, l, and with direction, k. These dependencies are contained within the particle fields; the particle fields were built to contain all relevant information. There is no reason to attribute a different scientific law to a different branch or to prefer any one direction on a branch to another direction or another branch. Hence, we expect that our constants should be universal.

The absence of directional preference in the equation implies that the universal coefficients, $\hat{c}_{\lambda}$, display a non-discriminating form. Mirroring Dirac's arguments, we shall let symmetry determine the values of these constants. Firstly as required by the separability of the dual division-ring symmetries, we note that the imaginary dimensions should exhibit the same intra-relationships as those possessed by the mathematical-real dimensions. Hence we represent $\hat{c}_{\lambda}$ as the components of a vector

$$\underline{\underline{c}}_1 = c_1\left(\left(1+\hat{\varepsilon}_1+\hat{\varepsilon}_2+\hat{\varepsilon}_3\right)+\hat{\beta}\left(1+\hat{\varepsilon}_1+\hat{\varepsilon}_2+\hat{\varepsilon}_3\right)\right). \qquad (10.83)$$

Secondly, without favouring any direction, each component of this vector should have the same size. Thus we must have

$$\hat{\varepsilon}_i^2 = \hat{\beta}^2 = 1 \qquad \text{for} \qquad i = 1\ldots3. \qquad (10.84)$$

Without loss of generality, we can rescale the equation of motion (with an associated renormalization of g)

$$g_1 \rightarrow c_1 g \qquad (10.85)$$

and thus we eliminate c_1 from the equation. Thirdly the form of the equation should be preserved under arbitrary coordinate rotation; hence the 3-vector branches are required to display quasi-vector anti-commutation symmetries

$$\hat{\varepsilon}_i\hat{\varepsilon}_j + \hat{\varepsilon}_j\hat{\varepsilon}_i = 0 \quad \text{for} \quad i \neq j . \tag{10.86}$$

An illustrative simple argument gives a further scientific explanation for this asymmetry. We shall consider the mathematical product of X and Y. We imagine that we exchange the X and Y coordinate directions and thus reverse the product that we are considering. Now, this coordinate exchange changes the parity of a subspace holding the X and Y directions. We cannot recover the coordinate relationships by rotation or displacement. However, we do regain the same coordinate system by applying a suitable rotation and reversal in either the X or Y coordinate direction. In other words, flipping the sign of the product is equivalent to reversing the order of multiplication.

This scientific argument applies to any plane. Thus fourthly we assert that the mathematical-imaginary directions anticommute with their complementary mathematical-real directions.

$$\hat{\beta}\hat{\varepsilon}_i + \hat{\varepsilon}_i\hat{\beta} = 0 . \tag{10.87}$$

Now, we write $\hat{\alpha}_i = \hat{\beta}\hat{\varepsilon}_i$ and easily confirm that the quantities $\hat{\alpha}$ and $\hat{\beta}$ satisfy the anti-commutation requirements of the Dirac matrices ie

$$\hat{\alpha}_i\hat{\alpha}_j = \hat{\beta}\hat{\varepsilon}_i\hat{\beta}\hat{\varepsilon}_j = \hat{\beta}\left(-\hat{\beta}\hat{\varepsilon}_i\right)\hat{\varepsilon}_j = \hat{\beta}\left(\hat{\beta}\hat{\varepsilon}_j\right)\hat{\varepsilon}_i = -\hat{\beta}\hat{\varepsilon}_j\hat{\beta}\hat{\varepsilon}_i = -\hat{\alpha}_j\hat{\alpha}_i \tag{10.88}$$

and

$$\hat{\beta}\hat{\alpha}_i + \hat{\alpha}_i\hat{\beta} = \hat{\beta}\left(\hat{\beta}\hat{\varepsilon}_i\right) + \left(\hat{\beta}\hat{\varepsilon}_i\right)\hat{\beta} = \hat{\varepsilon}_i - \hat{\beta}\left(\hat{\beta}\hat{\varepsilon}_i\right) = 0 . \tag{10.89}$$

To simplify the equation of motion and to transform the equation into a form that more closely resembles the Dirac equation, we employ the technique of separation of variables. We separate the dependence of $\underline{\underline{\phi}}^{(j)}(\tilde{q})$ on conjugate velocity $\{w_0, \overline{W}\}$ terms from its dependence upon non-velocity terms.

$$\underline{\underline{\phi}}^{(j)} = \underline{\underline{\psi}}(\tilde{q}_0, \tilde{q}_5, \tilde{q}_6, \tilde{q}_7)\underline{\underline{\zeta}}(\tilde{q}_4, \tilde{q}_1, \tilde{q}_2, \tilde{q}_3) \tag{10.90}$$

Hence, we obtain a pair of coupled equations. The first equation is

$$\begin{aligned} &\hat{\beta}\frac{\partial}{\partial\tilde{q}_4}\zeta_\upsilon^{(j)} + \left(\hat{\varepsilon}_1\frac{\partial}{\partial\tilde{q}_1} + \hat{\varepsilon}_2\frac{\partial}{\partial\tilde{q}_2} + \hat{\varepsilon}_3\frac{\partial}{\partial\tilde{q}_3}\right)\zeta_\upsilon^{(j)} \\ &\qquad - gB_{\upsilon\circ\lambda}^{(j)}\zeta_\lambda^{(j)} + C_S\zeta_\upsilon^{(j)} = 0 \end{aligned} \tag{10.91}$$

Multiplying this equation (10.91) though by $\hat{\beta}$ gives

$$\frac{\partial}{\partial\tilde{q}_4}\zeta_\upsilon^{(j)}+\left(\hat{\alpha}_1\frac{\partial}{\partial\tilde{q}_1}+\hat{\alpha}_2\frac{\partial}{\partial\tilde{q}_2}+\hat{\alpha}_3\frac{\partial}{\partial\tilde{q}_3}\right)\zeta_\upsilon^{(j)} \\ -\hat{\beta}gB_{\upsilon\sigma\lambda}^{(j)}\zeta_\lambda^{(j)}+\hat{\beta}C_S\zeta_\upsilon^{(j)}=0. \qquad (10.92)$$

Here C_S is a separation constant given by the second equation which is expressed in terms of the conjugate-velocity variables

$$\frac{\partial}{\partial\tilde{q}_0}\underline{\underline{\psi}}^{(j)}+\left(\hat{\alpha}_1\frac{\partial}{\partial\tilde{q}_5}+\hat{\alpha}_2\frac{\partial}{\partial\tilde{q}_6}+\hat{\alpha}_3\frac{\partial}{\partial\tilde{q}_7}\right)\underline{\underline{\psi}}^{(j)}=C_S\underline{\underline{\psi}}^{(j)}. \qquad (10.93)$$

As a constant of separation, C_S, plays several roles. It may be viewed as the source of rest mass in these equations. In a later section, it is viewed as a measure of the quantity of information associated with the velocity directions. In another sense, it embodies the influence of the rest of the universe upon the particle field. We understand this latter role by trying to estimate the constant, C_S. The following argument is intended to be shed light on what the various terms mean.

Physicists will be familiar with equation (10.92) as a linearized version of the Klein-Gordon equation; however, the counterpart equation (10.93) is less recognizable immediately. Thus we will insert a short digression to illustrate what this equation represents. This digression will draw upon some conventional knowledge (not derived from eventual reality) i.e. the digression will be technically non-rigorous.

Fourier complementarity identifies a position in coordinate space with the specific-momentum differentiation operator. Likewise, the associated quaternion-scalar differentiation should give the time interval, t, from an origin. The origin can be identified with the big bang as the only reliable reference location. The importance of this association will be explained more fully in a later section. Now, the separation interval from the big bang is fully time-like. Thus equation (10.93) implies that the constant, C_S, can be evaluated from

$$C_S=\frac{\partial}{\partial w_0}=\frac{\hat{i}m_0c^2t}{\hbar} \qquad (10.94)$$

Hence we discover that

$$-\hat{i}\hbar HC_S=m_0c^2HA_t \qquad (10.95)$$

where A_t is *the* age of the universe referenced to a reference frame in which the microwave background (our handle on the big bang[10]), positioned away from gravitational fields, has no directional red-shifts.

Consequently, whereas the differential equation for $\zeta^{(j)}$ corresponds to the classical relativistic relationship known as the Klein-Gordon equation, the differential equation for $\psi^{(j)}$ corresponds to the classical relativistic result

$$\left(\frac{m_0 c^2}{\hbar}\right)^2 (\tau^2 - \mathrm{X}^2/c^2) = \left(\frac{m_0 c^2 t}{\hbar}\right)^2 \qquad (10.96)$$

which with the identification $\mathrm{X} = v\tau$ corresponds to relativistic time dilation

$$t = \tau\sqrt{1 - v^2/c^2}\,. \qquad (10.97)$$

This is the end of the short digression to explain the physics of equation (10.93) and to estimate the value of the constant C_S.

Now, our spatial-coordinate equation-of-motion yields

$$\begin{aligned} -\frac{\partial}{H\partial t}\zeta_{\upsilon}^{(j)} &= \frac{c}{H}\left(\hat{\alpha}_1\frac{\partial}{\partial X_1} + \hat{\alpha}_2\frac{\partial}{\partial X_2} + \hat{\alpha}_3\frac{\partial}{\partial X_3}\right)\zeta_{\upsilon}^{(j)} \\ &\quad - \hat{\beta}\, g B_{\upsilon\circ\lambda}^{(j)}\zeta_{\lambda}^{(j)} + \hat{\beta}\, C_S \zeta_{\upsilon}^{(j)} \end{aligned} \qquad (10.98)$$

or using the estimate for C_S

$$\begin{aligned} \hat{i}\hbar\frac{\partial}{\partial t}\underline{\underline{\zeta}}^{(j)} &= -\hat{i}\hbar c\left(\hat{\alpha}_1\frac{\partial}{\partial X_1} + \hat{\alpha}_2\frac{\partial}{\partial X_2} + \hat{\alpha}_3\frac{\partial}{\partial X_3}\right)\underline{\underline{\zeta}}^{(j)} \\ &\quad + \hat{i}\hbar H g\,\underline{\underline{B}}^{(j)}\underline{\underline{\zeta}}^{(j)} + \hat{\beta} m_0 c^2 H A\,\underline{\underline{\zeta}}^{(j)}. \end{aligned} \qquad (10.99)$$

Thus this equation contains quantum-field interactions but no classical electromagnetic-field interactions. Otherwise, the equation would look like the Dirac equation provided the presupposed constant H becomes a variable equal to the inverse of the age of the universe. Several other reasons for accepting this assignment are provided later.

Since previously we did assume that H was a constant, we do need to comment upon the consequences of making H a variable. For any practical experimental time interval, the impact of laboratory-time passage on the value of H is unimportant

$$\begin{aligned} \Delta(H(t_1 - t_2)) = H\Delta(t_1 - t_2) + (t_1 - t_2)\Delta H \approx H\Delta(t_1 - t_2) \\ \text{for } (t_1 - t_2) \ll H^{-1}. \end{aligned} \qquad (10.100)$$

The important point is that we need to make clear from where we are measuring time. Whereas the age of the universe is referenced to the big bang, general time is to be referenced to a more recent time eg midnight, a calendar zero (AD), an instrument calibration or the astronomical present[11] (AP).

$$\frac{(t+\Delta t)_{re\ midnight}}{(t+\Delta t)_{re\ big\ bang}} \approx \frac{(t+\Delta t)_{re\ midnight}}{(t)_{re\ big\ bang}} \qquad (10.101)$$

Hence, for the most part, it is acceptable to treat H as a constant. However, both the Dirac equation and conservation of energy will need modification for investigations of the early universe or investigations involving cosmic-length time intervals.

In passing, I note that Dirac's α and β can be represented by the two-by-two matrices

$$\hat{\beta} = \begin{pmatrix} \hat{\gamma}_0 \hat{r} & 0 \\ 0 & -\hat{\gamma}_0 \hat{r} \end{pmatrix} \qquad \hat{\alpha}_1 = \begin{pmatrix} 0 & -\hat{\gamma}_1 \hat{r} \\ \hat{\gamma}_1 \hat{r} & 0 \end{pmatrix}$$
$$\hat{\alpha}_2 = \begin{pmatrix} 0 & -\hat{\gamma}_2 \hat{r} \\ \hat{\gamma}_2 \hat{r} & 0 \end{pmatrix} \qquad \hat{\alpha}_3 = \begin{pmatrix} 0 & -\hat{\gamma}_3 \hat{r} \\ \hat{\gamma}_3 \hat{r} & 0 \end{pmatrix}. \qquad (10.102)$$

This 2×2-matrix representation explicitly reveals the prediction (as provided by Dirac's equation) of the dual-spin-state character of all elementary fermion particles.

Finally we observe how the fields on different branches assemble into quantum states. We identify the products of the particle fields over the forward-time branches and over the backward-time branches with quantum-mechanical particle wavefunctions

$$\underline{\underline{\psi}}_{+}^{(j)} = \prod_{l=0..3} \underline{\underline{\zeta}}_{l}^{(j)} \text{ and } \underline{\underline{\psi}}_{-}^{(j)} = \prod_{l=4..7} \underline{\underline{\zeta}}_{l}^{(j)}. \qquad (10.103)$$

We take the particle field equation on each branch. We multiply by the fields on the other branches of the set and then add all four differential equations to find

$$\hat{i}\hbar \frac{\partial}{\partial t} \underline{\underline{\psi}}_{+}^{(j)} = \left[-\hat{i}\hbar c \left(\hat{\alpha}_1 \frac{\partial}{\partial X_1} + \hat{\alpha}_2 \frac{\partial}{\partial X_2} + \hat{\alpha}_3 \frac{\partial}{\partial X_3} \right) + 4\left(\hat{i}\hbar H g \underline{\underline{B}}^{(j)} + \hat{\beta} m_0 c^2 H A \right) \right] \underline{\underline{\psi}}_{+}^{(j)}. \qquad (10.104)$$

This form of the equation-of-motion is the form that resembles the Dirac equation most closely. Now, in the absence of external interactions with other particle fields, there can still be some self-energy contributions. Thus in the following chapter we will consider the rest mass to have two contributions, $m_1 + m_2$. The second contribution, m_2, is attributed to a self-energy term, which we explicit group with the identified mass term.

10.7 Bosons

We shall not pursue in detail a boson equation-of-motion here; rather we simply aim to show how the bosons are generated. Note that undressed boson fields can be extracted in a similar manner to that employed to identify the fermion-particle fields. As before, we seek a local constituency breakdown of the particle field into discrete superposable (ie linear) elements. Thus iteratively we shall introduce new fields satisfying linear equations. These new fields will progressively account for higher and higher powers of the particle fields; however, these new fields will all exhibit commuting behaviour.

Thus in the first iteration we introduce second-sequence fields, $\underline{\underline{B}}^{m,n,(j)}_{,P2}$,

$$B^{m_1 \circ m_2, n_1 \circ n_2, (j)}_{\lambda_1 \circ \lambda_2} B^{m_2, n_2, (j)}_{\lambda_2, P2} B^{m_1, n_1, (j)}_{\lambda_1} = \hat{r}\hat{\gamma}_0 \hat{e}_I \, . \qquad (10.105)$$

In the second iteration, we introduce third-sequence fields, $\underline{\underline{B}}^{(j)}_{,P3}$,

$$B^{m_1 \circ m_2, n_1 \circ n_2, (j)}_{\lambda_1 \circ \lambda_2, P2} B^{m_2, n_2, (j)}_{\lambda_2, P3} B^{m_1, n_1, (j)}_{\lambda_1, P2} = \hat{r}\hat{\gamma}_0 \hat{e}_I \qquad (10.106)$$

Explicitly the hybrid-space entropy parameter contains more field terms

$$S^{m,n}_{k,J} = \int d\tilde{q} \;\; \Lambda_{k,l}\left(\{\underline{\underline{\phi}}^{(j)}\}, \{\underline{\underline{\phi}}^{(j)}_{D\lambda}\}, \{\underline{\underline{\mathrm{B}}}^{(j)}\}, \{\underline{\underline{\mathrm{B}}}^{(j)}_{D\lambda}\}, \ldots\right). \qquad (10.107)$$

As previously the Lagrangian constraints deliver recognizable interactions with local-field coupling. Note that the fermion-particle fields couple to the dressed boson field corresponding to all the boson particles. Individual bosons couple to all other bosons (as generated by the iteration process) *and* couple to the undressed fermion-particle fields.

Note that, in most general form, we would decompose the particle field into odd-power and even-power terms. As required by the exclusion principle, the odd-power terms delivers only a single dressed fermion. Moreover, the even power terms deliver multiple boson fields.

10.8 Contravariance and Covariance

Finally, before we leave this discussion of Dirac's equation, let us make some remarks about covariance and contravariance.

In special relativity, it is common to adopt an incomplete tensorial phraseology that describes most 4-vectors as either only transforming covariantly (eg time-position) or only transforming contravariantly (eg energy-momentum and conjugate velocity). These contravariant and covariant vector styles are reciprocal conjugates eg a covariant position is conjugate to a contravariant momentum.

Now, all lengths are said to shrink in a moving inertial frame, whereas all momenta expand in a moving frame. Thus, in standard convention, we can generally pick whether a vector is to be covariant or contravariant by the way it scales in a moving frame. Such a relationship can inadvertently give the impression that the terms, covariance and contravariance, are descriptors of the how a variable scales in a moving frame.

The valid distinction between covariance and contravariance is easily demonstrated by considering two definitions for velocity 4-vectors. Clearly if time-position $(ct, \underline{x})$ transforms covariantly, then we can differentiate this 4-vector with respect to the invariant interval, s, (ie relativistic-proper time) and the result will be a covariant velocity

$$\left(c\frac{\partial t}{\partial s}, \frac{\partial t}{\partial s}\frac{\partial x_1}{\partial t}, \frac{\partial t}{\partial s}\frac{\partial x_2}{\partial t}, \frac{\partial t}{\partial s}\frac{\partial x_3}{\partial t}\right) = \gamma\left(c, v_x, v_y, v_z\right) \qquad (10.108)$$

where

$$\gamma = \frac{\partial t}{\partial s} = \frac{1}{\sqrt{1 - v^2/c^2}} \qquad v_i = \frac{\partial x_i}{\partial t}. \qquad (10.109)$$

Furthermore, since energy-momentum is contravariant, dividing this 4-vector by the rest mass, which is an invariant, will deliver a contravariant velocity. Using the standard expressions for energy and momentum

$$E = \gamma m_0 c^2 \qquad p_i = \gamma m_0 v_i \quad (i = 1..3). \qquad (10.110)$$

The contravariant velocity becomes

$$\left(E, -p_x c, -p_y c, -p_z c\right)/\, m_0 c = \gamma\left(c, -v_x, -v_y, -v_z\right). \qquad (10.111)$$

Note that both velocity 4-vectors are observed to scale the same way in a moving frame yet one 4-vector is covariant and the other 4-vector is contravariant. The sign difference between (the vector components of) these variables is like that difference between quaternion conjugates. The choice of positive or negative sign is tied to the variable's sign convention.

In eventual reality we always have both contravariant and covariant forms; however, this pairing of tensorial styles is provided by the dual spaces: coordinate space and quasi-space. Of course we can work with conjugates in the same space. When they are in different spaces, our tensorial conjugates obey different algebrae.

In the spaces, we have combined these two (position and velocity) 4-vectors into a single number (an 8-vector). Consequently we will choose to call the full hybrid-space 8-vector contravariant. Correspondingly, in coordinate space, the associated mathematical-complex quaternion vector representing both position and conjugate velocity is necessarily referred to as covariant. However, the positive sign convention is attributed to contravariant variables in hybrid space and to covariant variables in coordinate space.

10.9 The Measurement Side of Quantum Mechanics

Now, quantum mechanics is often thought of as being a non-local theory because quantum mechanics does recognize some non-local attributes in nature. However, we have seen that quantum mechanics retains many inherited[12] *local* features such as a presumed particle-field basis. Thus quantum mechanics may be regarded as providing a bridge between a non-local aether in coordinate space and the definite rigidity of a completely local classical mechanics. Quantum measurement and the quantum equation-of-motion are separate spans of this bridge.

Quantum mechanics generates many measurement-related dilemmas that cannot be answered from within the belief structure of quantum mechanics. Why are there two distinct and incompatible parts that we identified as the principles of quantum mechanics and the rules of quantum measurement? Why are the results of measurement classical variables? Why does quantum reality provide probabilities for experimental outcomes? Why are probabilities constructed as the square of the (mathematical-complex) modulus of wavefunctions? What constitutes a physical measurement? When is the measurement completed? How does a locally determined experimental result propagate a change of state throughout the entire studied system? Why is measurement irreversible? Eventual reality provides a picture that can answer such questions. In this section, we shall discuss key aspects of eventual reality's view of quantum measurement.

Eventual reality recognizes a hierarchy of approximate theoretical representations of reality. In this hierarchy, the principles of quantum mechanics sit on a storey that is different from the storey on which the measurement rules reside. The rules of measurement are more classical (and definite) than the quantum principles, their conceptual outlook is different and, consequently these measurement rules are positioned in the hierarchy further away from the fundamental storey of eventual reality. Indeed, when the founders of quantum mechanics tethered the Copenhagen interpretation of quantum reality to classical measurable observable phenomena and quantities, they were actually implicitly recognizing that classical reality is linked to these (quantum) measurement rules.

The hierarchy of scientific theories is invented to make sense of perceived reality. At a fundamental level, everything is (ie is possible, exists, happens, etc). What we recognize as actions, sense and order come about as correlations within this anarchy. Our observations of the spatial correlations within the universe are myopically selected to be comprehensible by an individual localized awareness. Thus describing certain local correlations within the wider eventual reality delivers the (partially local) principles of quantum mechanics. Now, we will explain the characteristic features of the next storey (ie the measurement storey). The scope of science covers only what we can objectively predict. Hence science has been a study of objective cause and effect. Consequently it is not surprising that historically we should synthesize a theory of quantum measurement that can attribute an absolute certainty to the result. We shall discuss how partially adding this feature generates the measurement level in the hierarchy of theories.

Thus quantum measurement is positioned between the principles of quantum mechanics and classical reality. Fully adopting a deterministic certainty generates the principles of classical mechanics that require that every potentially measurable quantity has a precise knowable value ie all obervables possess a permanent unique definite objective existence. The attractiveness of this classical philosophical dogma involving certainty, definiteness and predictability has provoked and plagued science for centuries. Only eventual reality finally lays this dogma to rest by describing the detail of a rational fundamental underworld from which quantum mechanics is derived.

Next, we shall discuss measurement from two different standpoints: one mechanistic and the other procedural.

10.9.1 The Mechanisms of Measurement

Measurement is a process in which selected information is apparently transferred from the study system to a measurement apparatus in an irreversible way which puts the study system into a new state as identified by the measurement result.

All measurement systems exhibit some classical-style behaviours. Thus these aspects of the measurement system can be regarded as being so complex that the details of their microstates are unimportant.

The macro states & variables of the measurement apparatus satisfy the everyday classical physical laws and principles including determinism, causality, distinctness (unique identifiable local presence), continuity and smoothness of available values of analogue macro-variables, robustness (no change for no reason) and time-reversal invariance.

From a quantum perspective, consider the interaction of a simple quantum state studied with a macroscopic complex measurement system possessing robust classical behaviour. In many experiments, we would compute using the boundary conditions that some of the simple quantum state would be spatially reflected with reversed parity. However, a traditional quantum analysis would not include a similar time-reversed reflected state.

From the perspective of eventual reality, our measurement system is a bound localized many-particle system that is heavily populated and linked across all branches. This measurement system is insensitive to the branch of an incident quantum state. Indeed, the response of the measurement system to the incident quantum state is practically uniform across the branches and the response can be imagined as acting like a partial short-circuit across these branches of space. This linkage of branches is across all eight branches ie it includes those branches with a different parity and those branches with a different time-sense. Thus we retain systemic behavioural similarities in space and time. Interactions are local and cause a change in direction in eight dimensions and (for interactions with a measurement apparatus) in eight branches. In other words, a change of branch is simply a reflection.

It has long been recognized[13] that allowing retro-causal time-reversal reflections will permit relativistic transformations of EPR-style measurement experiments. Many experiments include an initially set-up (this procedure is like a time-reversed measurement) followed by one or more measurements. The set-up phase may correlate some components of the study-system; subsequently these components spawn subsystems that can be measured separately. In the causal Copenhagen interpretation, a measurement instantly changes a system of spatially separated subsystems in a blatant denial of relativistic principles (the EPR paradox).

However, eventual reality accepts and even requires both a partially reflected backwards-in-time signal[14] at the measurement equipment and a signal that is partially reflected forwards in time by the relevant set-up system. This loop of reflections (the future influencing the past and the past influencing the future) shuffles the potential experimental outcomes precluding absolute determinism. Ultimately the physical measurement delivers an unpredictable result that is selected in keeping with outcome probabilities ie the product of probabilities across all eight branches.

The filter is another useful descriptive analogy of the interaction mediating a measurement. A measurement system may be viewed as a filter that only lets through the final measured result, reflecting backwards in time the other rejected components. Then, this act of measurement is seen to directly affect both the past (back to a state set-up) and the future (forward to the next state set-up or measurement). Intent is often exhibited in these set-ups thereby limiting the further backward flow on of consequence from this simple act of measurement.

10.9.2 The Character of Measurement

Measurement requires an observed system, a measurement instrument and an interaction between the observed system and the measurement system. We shall restrict the considerations here to quantum observable systems rather than systems which are well-described by classical mechanics. The measurement instrument is required to display an element of classical reality that interacts with the quantum observable system.

The main features of this interaction form the ABC of measurement. These three features are Amplification, Branch blindness and Certainty.

10.9.2.1 Amplification

The primary reason why we are considering only quantum observable systems is that measurement of a quantum system necessarily involves amplification of the information being measured. Measurement on a system that is already exhibiting classical behaviour does not require such amplification (and does not require the quantum rules of measurement). By amplification, we are referring to an aspect of the operation of the measurement apparatus that generally detects a quantum quantity, maps this quantity onto a particular measurement signal that is amplified into a classically observable quantity. However, this measurement amplification is always a two-way street.

Consider the classical example of a high-gain mechanical lever. A small movement of the detection end of the lever produces a large movement at the measurement end. However, in turn, a small force at the measurement end of the lever produces a large force at the detection end. Thus the lever amplifies any random force fluctuations coming from the measurement side creating uncertainty in the localized forces at the 'point' of measurement. This analogy exhibits the concepts of the pairing of complementary variables and of two-way amplification. These concepts are features of all quantum measurements. Newton explained that for every influence there is an equal and opposite reaction. In the presence of amplification, this reverse influence is felt through the disturbance of the complementary variable. Indeed, this condition is a restatement of Heisenberg's uncertainty principle. The accurate measurement of one variable is accompanied by a mammoth disturbance in its complementary variable. This classically valid result (Heisenberg's uncertainty principle) provides an interpretation as to why the classical world is represented by mathematical-real quantities. A perfect measurement only ever copies half an answer, measurement unrecognizably changing the other half of the answer. Quite generally the mathematical real and imaginary components of a quantum-system variable are measurement complementary. Amplification stabilizes the value being measured and scrambles its mathematical-complex complement. These measurements, delivering mathematical-real results, have far-reaching consequences that pervade experience and shape our common sense notion of what constitutes reality.

10.9.2.2 Branch Blindness

Measurement is blind to the branch structures of eventual reality's spaces. The enforced mathematical-real classical-style reality disguises some of nature's honest symmetries. Measurement can only sample and amplify one aspect of the system under study and it will disrupt another complementary aspect.

From the vantage point of eventual reality's worldview, the quantum mechanical equation-of-motion provides a branch wavefunction that, in measurement, we equate to a probability (not probability-amplitude) density. A superposition, of the actualities on the eight branches of space, provides our branchless classical reality and it provides the measurement results. Thus the classically mathematical-real probability distribution is simply the product of these eight individual branch probability densities. However, in temporally sensitive experiments, it is appropriate to gather the branches into the forwards-in-time branches and the backwards-in-time branches. The combined probability density for forwards-in-time branches is the conventional wavefunction of quantum mechanics. The combined probability density for backwards-in-time branches is the mathematical-complex conjugate of this conventional wavefunction. The mathematical-real-centric background of quantum measurement leads to the quantum-mechanical interpretation of the wavefunction as a mathematical-complex probability-amplitude density.

10.9.2.3 Certainty

The measurement instrument contains components that belong to the realm of classical mechanics. These classical components are required to display data but also they can interact with the system to be measured.

Measurements have a definite outcome. Simply the statistical distributions on the different branches are co-dependent. Just as the statistics of the eight directions provide linked (Fourier) complementary four-variables, the eight branch directions generate linked symmetry-displaying statistics across the branches. Symmetry is the provider of all structure and correlation; the set of branch wavefunctions must mirror the symmetry of the branch set. A mathematical-complex conjugate will reflect a time reversal; a quaternion conjugate will reflect a parity reversal. Then, multiplying such branch wavefunctions yields a certain definite (mathematical-real quaternion-scalar) result for a measurement.

10.9.3 CPT Symmetries

Finally we will discuss how the relationship between the charge conjugation (C), parity (P) and time sense (T) symmetries arise in measurement. Thus let us define these sign-flipping operators as

$$P \equiv (-)_P \,,\; T \equiv (-)_T \,,\; C \equiv (-)_C (-)_B \,. \qquad (10.112)$$

Now, there is a similarity between a particle travelling forward in left-parity space and a particle travelling backwards in right-parity space. Consider the sign $(-)_C$ applied to each of the eight branches. When we reorder the branches we observe that we have switched the parity. Now, consider applying the sign $(-)_P$, this sign reverses the directions of space ie switches parity. Likewise, we associate time-direction reversal $(-)_T$ and time-sense-branch switching $(-)_B$. The mathematical signs of each pair are similar but not equivalent. Nevertheless, during a measurement, the branches have been 'shorted' together (as expressed by boundary conditions) so that the signs have a common impact on a space of compacted branches. Within the branch-compacted space (coinciding with a measurement), we observe that applying

$$PCT \equiv (-)_P (-)_C \; (-)_T (-)_B \qquad (10.113)$$

has no impact on quantum-measurement results. However, the action of a PCT operation can change what happens on individual branches. Furthermore, we cannot automatically expect just one or two of these sign-flipping operators to leave a system measurement unchanged; we require all three operators for the signs to effectively cancel.

10.10 Summarizing the Interpretation of Quantum Mechanics

Eventual reality explains much more than quantum mechanics. As demonstrated in this document, eventual reality can describe space and time, mind and matter and a sweep of fundamental principles including analyticity, determinism, causality, equivalence, correspondence, etc. Here we shall summarize how the interpretation of quantum mechanics differs from the Copenhagen interpretation.

In eventual reality's interpretation of quantum mechanics, the state of the system is real; the branch-based wavefunction is taken to be a form fully describing a physical system. The individual particles states are part of a decomposition of a representation of the order in the aethereal field.

However, the wavefunction ceases to be a probability-style entropy parameter and the wavefunction becomes the information description of the reified system. Thus the physical reality of the wavefunction reinterprets this entropy as information. Furthermore, this information is the content of the material world of experience. The chain of reasoning simply is: the quantitative world is described by information; information is the wavefunction; the wavefunction is a numerical measure of an underpinning qualitative reality.

All interactions occur in space and time and the causal consequences are scattered in space and time (forwards and backwards). Measurement involves three phases: an initial phase in which the system is set up, an intermediate phase and a final phase in which an observation is made on the system. From the perspective of a reference frame, the system at times during the intermediate phase is contingent; the system only becomes assigned at times beyond the current observation. Thus the description of state changes occurring in measurement, transforms in a consistently relativist manner. The observation phase involves the coupling of the quantum system to classical reality as exemplified by a lack of information differentiation across the branches of space-time and a statistical maintenance of macroscopic ensemble form. A flowing river provides a familiar example, the water is continually changing and moving yet the ensemble form, which we call the river, is maintained as a relatively fixed structural form. Likewise, the fundamental constituent parts of a classical variable can be fickle yet the assembled multi-constituent form maintains a definite semi-permanent consistency.

10.10.1 The Quintessential Quantum Questions

Quantum mechanics, as described by the Copenhagen interpretation, is full of mystery. Eight questions usefully highlight the seemingly inexplicable peculiarities of quantum mechanics. We shall now summarize the answers, to these questions, provided by eventual reality.

1. *Why does quantum mechanics relate to experiment via probabilities?*
There is an underpinning seemingly chaotic foundation layer of reality. This foundation is known as the apeiron and it is a realm of unlimited opportunity. From the apeiron we derive a quantitative coordinate space which is filled with a qualitative aether. The aether holds the description of everything (material and mental) existing within space; we use a gauge of statistical order to extract and describe a quantitative element of reality. Thus quantum mechanics is derived from a statistical theory. Probabilities are natural in such theories.

2. *Why is matter a bundle of discrete quanta?*
The quantum transformation of the collective features of the aether are analysed in terms of linear discrete contributions known as particles of matter. This analysis is useful since it creates a phenomenological rich description of the aether. This is why we have constructed the edifice of reality in this form.

3. *Can we apply quantum mechanics to the entire universe?*
We only observe matter in hybrid space. Hence the equations of quantum mechanics are derived as a transformation of a finite region of the foundation coordinate space. These transformations cannot be applied to all of coordinate space; however, the equations of quantum mechanics may be applied to all of observable (hybrid) space.

4. *Is quantum mechanics irretrievably in disagreement with relativity?*
Actually the disagreement is an illusion. Special relativity finds the instantaneous propagation of consequence across entangled states challenging. However, this instantaneous effect becomes relativistically transformable when we allow measurements to produce consequences that spread both forwards and backwards in time. Compatibility with general relativity is often thought to be a bigger problem. However, quantum mechanics describes the dynamics of matter (in flat space) and general relativity describes the topology of space (for classical matter). We shall explain in part 3 that the two theories are different approximations to one and the same equation; both are derived as approximations to the same first principle of the universe.

5. *What is responsible for the decoherence that is involved when a quantum macroscopic superposition adopts a classical value?*
Every quantum state including an entangled state (eg the live/dead state of Schrödinger's cat) will have intermittent minor interactions with the environment. These minor interactions will cause repeated minor reflections of the quantum state; the accruing reflected contributions will be redirected by the registration of the system's initial set up. Over time these minor reflections will cumulatively steer the quantum state into one with an environmentally stable (ie classical) outcome.

6. *Why are there two distinct and incompatible parts that we identified as the principles of quantum mechanics and the rules of quantum measurement?*
The quantum equation-of-motion describes the quantitative aspects on the second-elevated storey of reality. Measurement does not enter into the precepts of this storey. Measurement is a natural consequence of the third-elevated storey, which permits the embedding of localized classical realism in a wider quantum system. Thus the clear distinction is a reflection of the formalisms residing on different levels of reality.

7. *Why are the results of measurement classical variables?*
The results are mathematical-real numbers because we multiply the probabilities across all branches and the reflection boundary conditions ensure that the non-scalar non-real terms vanish. The results have classical definiteness by virtue of the weight of statistics. Toss a coin twice and there's a 50% chance that close to half the results are tails. Toss an unbiased coin a trillion times and you are practically guaranteed that (percentage wise) very close to half the results will be tails.

8. *What constitutes a measurement?*
Measurement is not like a make or break switch. The course of measurement is a gradual process moving from start to completion and mediated by interactions. It is convenient to picture every interaction as spawning other quantum state components with potentially different parities and time senses. Each interaction contributes to the gradual completion of the measurement.

Notes

[1] We shall not discuss this range until section 17.4.1.

[2] Every option is expressed in the foundation storey of reality.

[3] Conscious actions are an essential part of the hypothesis of eventual reality. However, in this chapter we only consider physical reality.

[4] The author hopes that the subsequent use of some tensor notation in part 3 does not provide problems for the reader.

[5] Collective coordinates are used to describe collective excitations of many-body systems, such as phonons and plasmons. The collective coordinates, see Percus and Yevick (1958), are well suited to disorganized systems and their expression is Fourier-like.

[6] For a definition of a left-module see Iyanaga & Kawada (1977) §§275C&D.

[7] The physical significance of associativity was discussed in section 9.2.2.

[8] For a definition of a Lie algebra see Iyanaga & Kawada (1977) §247A.

[9] Similarly, a Clifford algebra forms a unitary (non-division) ring. For an introduction to this algebra see Penrose (2004) p208.

[10] More correctly the cosmic ray background is the last remnant of radiation from the opaque plasma era of the early universe. As this plasma cooled, neutral matter formed and the universe became transparent.

[11] Astronomers dislike the inconvenience of a constantly changing present time and prefer to change the 'astronomical present' every 50 years.

[12] Quantum mechanics is traditionally derived from classical mechanics. These features are inherited from classical mechanics.

[13] Leggett (1977) discusses how a causal connection between independent measurements on entangled parts of a system (eg an EPR experiment) would need a time reversal of causality to be compatible with relativistic principles.

[14] See also the discussion in section 2.5.

Part 2

Chapter 11 : Comparison with the Tangible Universe

The hypothesis of eventual reality maintains that it is the frenetic acts of random choice exhibited by pointer links, happening on the founding primal level of reality, that creates calibrates and controls those aspects of the universe that are free of intent. We are to repeatedly use Fermat's principle to distil our perceived order from such perpetual pandemonium.

This chapter discusses how the details of the hypothesis of eventual reality most likely impact upon our physically tangible universe. For the purpose of generating this wider picture, this chapter will employ estimation and some known universal facts. Note that the chapter does not include ab-initio derivations of results from the tenets of eventual reality.

In part 2, we are presenting the first embryonic form of eventual reality. Recall that quantum mechanics has taken many forms over the last century. This embryonic form of eventual reality has not accommodated the principles of general relativity and gravity. However, both general relativity and eventual reality are relativistic topological theories and thus these theories possess no obvious philosophical impediments to being combined. The current embryonic form of eventual reality has enabled eventual reality to be rapidly explored and understood. However, the inclusion of general relativistic principles is likely to modify some quantitative prediction of this section. We should be particularly concerned with large-scale (cosmological) results. A discussion on the inclusion of spatial curvature is provided in part 3. However, a completed cosmology is not presented.

Now, coordinate space is built from two symmetries. The quaternion symmetry provides the structure for a recognizable non-trivial (1+3) dimensional space. As a potential *positional* space, the complex symmetry is less interesting, as it would generate a trivial (1+1) dimensional arrangement. Thus only the quaternion structure provides a viable physical space for intricate phenomena and the mathematical-complex symmetry group delivers the capabilities of an informational repository structure that can be possessed by a four-dimensional space. This dual structure helps to specify the principal degree of informational interdependence that provides that most obtrusive of spatial properties namely generic matter.

The universe, as described by eventual reality, is built from information encoded in the links between events. Thus matter is information encoded in space. This association is the reason for the connection in quantum mechanics, between statistics (described by possibility and probability) and matter. In general relativity, this association is the reason for the relationship that black holes have with entropy and temperature. This informational structure is quantifiable as a size in phase space – to be explored in section 11.3. This association is the reason for the connection in statistical mechanics between information (described by order or knowledge) and phase-space volume.

There is interplay between the big and the little in the universe. This interplay fixes aspects of our physical reality. Thus many laws that we hold to be true (and even some seemingly self-evident definitions) are asserted (by eventual reality) to be the net result of simple chaotic behaviours. We shall now expose some of these results relating to momenta, mass and information.

11.1 Momenta

Firstly we shall show that momentum is associated with the product of mass and velocity. Let us start with the concept of a particle trajectory, which has its origins in classical mechanics. Now, Newton's first law of motion requires that isolated free particles have straight-line trajectories. This classical result is described by the equation

$$\Delta X = \beta c \Delta T \tag{11.1}$$

Here, ΔX is the distance along the trajectory, ΔT is the time difference and β represents the speed of motion as a fraction of light speed, c. In wave theories, a uniform straight-line trajectory is readily understood as being formed from the superposition of a phase-correlated line of Huygens' wavelet sources.

In eventual reality, we interpret this result as an absence of choice. Choices are somewhat like interactions in that potentially they lead to recognizable changes. Thus not surprisingly it is useful to analyse choices in a similar fashion to the way we diagrammatically break-up multiple interactions in quantum field theory. Hence, we may choose to represent the cascade of choices in terms of all possible ways of assigning all choices separated by sections of no choice. It is instructive to consider what is meant by no choice. By 'no choice', we do not mean indecisive or random choices. No choices on offer would mean that the system's evolution is strictly determined and hence trajectories are analytically computable. Let us consider an elemental particle. Between choices, this particle would have an analytical trajectory, $q_\lambda(q_4\hat{i}\hat{\gamma}_0\hat{e}_V)^1$. However, the square of $q_4\hat{i}\hat{\gamma}_0\hat{e}_V$ is zero. Hence, the determined analytical expansion of $q_\lambda(q_4\hat{i}\hat{\gamma}_0\hat{e}_V)$ in powers of q_4 is linear. Thus the no-choice trajectory is given by the expression

$$q_\lambda(q_4 + \Delta q_4) = q_\lambda(q_4) + \left.\frac{\partial q_\lambda}{\partial q_4}\right|_{q_4} \Delta q_4 \tag{11.2}$$

ie for a choice-free spatial trajectory

$$\Delta X = \beta c \Delta T . \qquad (11.3)$$

Factorability of coordinate space implies that our equations should be formed from mathematical-complex analytical functions. This analyticity implies that the isolated classical trajectory should take the form

$$\left(\hat{r}\frac{H\Delta X}{c} - \hat{i}\Delta W_x\right) = \beta\left(\hat{r}H\Delta T - \hat{i}\Delta W_0\right) \qquad (11.4)$$

where $(\Delta W_0, \Delta \underline{W})$ is the change in the conjugate velocity. In this example, the speed, β, is taken to be a constant (in time) and hence the changes ΔW_λ are associated with changes in mass!

In a wave picture, the coherent relationship described by the two additional terms can also be understood as arising from the interference of Huygens' wavelets propagating in the imaginary dimensions. The additional terms yield the independent coherent result

$$\Delta W_x = \beta \Delta W_0 \qquad (11.5)$$

for the free motion of a particle. Multiplying through by the rest mass and speed of light effectively gives the Newtonian definition of momentum

$$\Delta P = \left(\Delta E / c^2\right)\beta c = \Delta m v . \qquad (11.6)$$

Thus factorability dictates that on the large-scale, momentum agrees with its classical definition ie momentum equates to the product of mass times velocity. Alternatively it shows the conjugate velocity, $P / m_0 c$, is derivable from $v = dX / dT$. These macroscopic-limit agreements can be compared to the requirement for energy conservation. In quantum mechanics, we may talk hypothetically about borrowing energy for a short period so long as the energy loan is repaid in full. The consequences are the same in eventual reality but the language is changed. We say that fluctuations in energy are present but that they become unimportant when we *average* over long periods of time. This argument is importantly different from saying that energy is conserved *after* a large period of time. Likewise, these macroscopic-limit results imply that momentum can deviate from the product of mass and velocity in small regions of space. However, the difference, between momentum and the product of mass and velocity, shrinks to zero as we *average* over larger and larger regions of space.

Thus we confirm the classical identification of velocity with the rate of change of location. In quantum mechanics, we represent the momentum (to within an invariant proportionality factor) as the associated mathematical-complex complementary variable to displacement. This representation becomes the definition of momentum in hybrid space. However, momentum in hybrid space can be related to momentum in coordinate space.

Physically the argument is simple. To the extent that the large-scale classical average result can be considered to hold in hybrid space

$$W_x = \frac{W_0}{c}\frac{\partial X}{\partial T} \qquad (11.7)$$

then, as in classical mechanics, W_x and ΔX are expected to be Fourier conjugates of each other. However, by definition, w_x and ΔX are Fourier conjugates as also are W_x and Δx. Fourier transforms are symbolically unique; hence physically we argue for the large-scale equivalences

$$< W_x > \doteqdot < w_x > \qquad (11.8)$$

and

$$< \Delta X > \doteqdot < \Delta x > . \qquad (11.9)$$

There are several subtleties that we have glossed over. For instance, the correspondence across different spaces does not require that the units must map linearly. Secondly we are strictly only discussing small fluctuations. Thirdly there is not any matter in coordinate space only aether.

Note that the origin of coordinate space is preserved in hybrid space for the dimensions represented by conjugate 3-velocity (specific momentum) and by time and energy (untransformed); however, an absolute reference for 3-position is lost in hybrid space.

11.2 Hubble Expansion

Our coordinate-system scale contains two dimensioned parameters: a comparative temporal scale, H, and a speed, c. These two parameters fix the relative sizes of the mathematical-real and mathematical-imaginary terms and the relative sizes of the quaternion-scalar and quaternion-vector terms.

$$\underline{\underline{\chi}} = \left(\hat{r}\, w_0 + \hat{i}\; Ht\right)\hat{\gamma}_0 + \left(\hat{r}\, w_1 + \hat{i}\; Hx_1 / c\right)\hat{\gamma}_1 + \left(\hat{r}\, w_2 + \hat{i}\; Hx_2 / c\right)\hat{\gamma}_2 + \left(\hat{r}\, w_3 + \hat{i}\; Hx_3 / c\right)\hat{\gamma}_3 \qquad (11.10)$$

c is naturally identified with the speed of light. Recall that Planck's constant, h, was introduced (in the reciprocal transform) to give the scale of phase-space. Both h and c can be taken to be constants calibrated by our Earthly unit system.

Choices of unit systems will abound if just one accepted constant of nature actually varies (presumably in a regular manner) with time. For consistency, we should require constancy against variations not only in time but also in space and conjugate velocity. Now, there is a natural-unit system in which the dimensionless size that is attributed to each linear dimension of an event, $Đ$, is constant. In these natural units, we might also expect that the classical scope of the eight dimensions would balance. The classical limit on 3-

velocity is the speed of light; thus the scope of the vector components of conjugate velocity is unity. Likewise, the classical scope of accessible 3-distance is the product age of the universe and the speed of light. Imposing this same scope on the dimensionless mathematical-imaginary vector components implies that H is a variable equal to the inverse of the age of the universe. Note that this identification also produces unit values for both the mathematical-real and the mathematical-imaginary components of the relativistic-proper dimensionless scalar coordinate. In this natural dimensionless unit system, the change of time (since the big bang) is exactly matched by a cancelling change in H.

A linear motion, in mathematical-complex phase-space, lineally expands or shrinks 4-space. Hence we conclude that our spatial universe displays a uniform expansion at a Hubble expansion rate of H.

The conclusion, that H is a variable, has implications for the gravitational constant. At this stage in our explanation of eventual reality, gravity has not been incorporated into the analysis. However, we should have every expectation that the Planck distance (as provided by classical and quantum mechanics) will provide the approximate distance scale to describe an event. This association will be confirmed in part 3.

The dimensionless Planck-distance scale will be taken to be given by

$$Đ \approx \frac{H}{c}\sqrt{\frac{3G\hbar}{2c^3}}. \qquad (11.11)$$

This expectation (that the scale of an event should equal the Planck scale) only justifies this equality (11.11) to within an order of magnitude. Note that the numerical factor in this equation has been doctored so as to engineer a later precise agreement that otherwise would only be approximate.

Finally observe that the gravitational constant, G, needs to vary with the square of the age of the universe if $Đ$ is to remain constant. However, we show later that masses vary inversely with the age of the universe. Thus gravitational strengths remain age independent. The relative smallness of present day, G, is associated with the smallness of the Planck distance compared with the size of the universe. Qualitatively this is a necessary condition for a multistorey edifice of macroscopic coherent description.

11.3 Information Structures

Phase space is the restricted 6-dimensional vector-space part of the full space. Matter arises in the transformation of this 6-space. Thus by specifying positions and momenta (or 3-vector conjugate velocities), we may define a persistent material system that can retain and display knowledge for an extended period of time. Some persistent material systems would be dynamically stable eg an orbit; other persistent systems could seem to be unchanging. Ultimately the system's useful behavioural knowledge is retained within the SU(2) subspace.

Consider any region of phase space. If we displace this region in a position direction the information capacity of this region should not change. Thus without a constraint to make a direction special, quite generally information capacity should be invariant under any phase-space displacement (ie position or conjugate-velocity directions). Now, we can add and subtract the measure of local information content. Thus, by adding infinitesimal regions together, we readily deduce that information capacity is an extensive property of phase space.

Hence the volume of phase-space would appear to give a measure of the count of useable information. The scale for calibrating this information is provided by Planck's constant. Historically Planck's constant was introduced as the scale of all quantizations. The constant appears in eventual reality via the (Fourier conjugate) correspondence between position and momentum.

Now, elemental information can be encoded across multiple levels in the aethereal hierarchy. Consequently from the uncertainty principle[2], we write the *dimensionless* elemental volume of phase space as h_0^3 where

$$h_0 = N\frac{\hbar H}{2m_1c^2}. \qquad (11.12)$$

Here the factor N (>1) represents the degeneracy provided by allocating information across the branches and directions of the higher-order structures in space. The values that N may take will be considered later. However, we can register this information another way. From a microscopic standpoint the unit of information is encoded in the event. Ultimately an event's pointer decodes a unit translation with no directional sense. Thus the elemental phase-space measure of the information content of single pointing event is just $Đ^2$.

Hence equating the large and small information-capacity scales gives

$$\left(\frac{H}{c}\sqrt{\frac{3G\hbar}{2c^3}}\right)^2 = \left(N\frac{\hbar H}{2m_1c^2}\right)^3. \qquad (11.13)$$

Thus we conclude that the m_1 contribution to rest mass is quantized

$$m_1 = Nm_{_{II}} \qquad (11.14)$$

in units of mass, $m_{_{II}}$, given by

$$m_{_{II}} = \left(\frac{3\hbar^2 H}{16Gc}\right)^{\frac{1}{3}}. \qquad (11.15)$$

Note that this mass-scale unit[3] varies inversely with the age of the universe. The rest masses of elementary particles will scale with this unit but (in general) the rest masses are not integer multiples of m_u. A second self-energy term also contributes to the measured rest mass. For other quantized elementary-particle properties one needs to be cognizant of further degrees of freedom of the universe. We shall observe that their quantum numbers do need to be describable only in terms of the dominant symmetries of the universe.

Let us consider an alternative physical explanation of the meaning[4] of this quantized mass-scale value. For the time being, we shall assume that mass is identified with the quantity of information, which is digitized in bits of information. We shall show later how interactions modify this simple association. The quantum unit of mass is to be an unknown, m_u. We want to be able to spatially localize (and separate) this information sufficiently to identify it with an individual unit. From the uncertainty principle, a registrable energy unit, $m_u c^2$, can only be squeezed into a spatial width of dimension $\hbar / 2m_u c$. Any further squashing will inevitably create extra unseparated units of information. Next, we estimate the greatest gravitational self-energy of this data unit as $\dfrac{4Gm_u^2}{3\hbar/(2m_u c)}$. In a universal sense, this self-energy must be detectable – we should deny relevance to anything that is undetectable. This self-energy represents the coupling of the particle to the space distortion that it has created. Indeed, (as the limiting self-energy of a single unit of quantum information) the maximum self-energy is expected to equal the minimum amount of energy that is measurable in the universe. This condition defines our digital count of information. If this energy were not measurable, the information could not be able to form a localizable recognizable sustainable imperfection (or distortion) in space: there would be no particles of information. Now, in the time interval since the big bang, the least quantity of energy that can be measured is $\tfrac{1}{2}\hbar H$. Thus equating this measurable least energy with the gravitational self-energy provides the mass quantum for the unit of information that can be embedded in space, (as previously) we get

$$\frac{4Gm_u^2}{3\hbar/(2m_u c)} = \tfrac{1}{2}\hbar H \cdot \qquad (11.16)$$

This equation simplifies to the previously obtained value of m_u (equation (11.15)).

11.3.1 Time-dependent Units

When our system of units is time dependent, then the characteristics of reality can vary between different unit systems. Consider the dimensionless time coordinate Ht. Both H and t need to be evaluated in common time units to calculate this product. Let us denote a general variable, x, as the product of an appropriate dimensional unit, ${}_x\breve{u}$, and a corresponding dimensionless size, x_u. Then, for two competing unit systems, we shall write

$$t = t_{u_1}\,{}_t\breve{u}_1 = t_{u_2}\,{}_t\breve{u}_2\,. \quad (11.17)$$

We shall derive a specific example to demonstrate how the picture of reality can be made to change by changing the units. For the competing systems of units, we shall build the basic dimensional units from fundamental constants[5] c, $\hbar$ and $\alpha_{1/3}$ supplemented with a dimensional mass unit.

The universe expresses information in mass units, m_u. This mass is directly proportional to the masses of elementary particles and thereby to conventional atomic units and to observational techniques. Thus m_u is chosen to provide the mass unit for system 1. Now, we have already assigned the Planck scale to the event. This Planck scale provides a mass unit of $\sqrt{\hbar c/G}$. Hence, the Planck mass will be taken as the mass unit for system 2. Clearly this second unit system is a useful system for describing the superposed networks of events. A comparison of the SI unit dimensions for these two suggested systems is given in table 11.1. Note that we shall use charge rather than current as the fourth unit dimension.

Table 11.1

The SI unit system nominally recognizes four basic dimensions of units (angles, temperature and Avogadro's number are dimensionless as are various dB scales and luminous intensity referenced to a standard candle).

Units of	System 1: Atomic	System 2: Planck
Mass	m_u	$\sqrt{\hbar c/G}$
Length	$\hbar/(m_u c)$	$\sqrt{G\hbar/c^3}$
Time	$\hbar/(m_u c^2)$	$\sqrt{G\hbar/c^5}$
Charge	$\sqrt{36\pi\alpha_{1/3}\varepsilon_0\hbar c}$	$\sqrt{36\pi\alpha_{1/3}\varepsilon_0\hbar c}$

Rewriting equation (11.15) for m_u, we get

$$\frac{16}{3}\,{}_t\breve{u}_2^2 = {}_t\breve{u}_1^3 H \cdot \qquad (11.18)$$

Now, writing H as the inverse of the age of the universe, A_t,

$$H^{-1} = \underset{u_1}{A_t}\,{}_t\breve{u}_1 \cdot \qquad (11.19)$$

Thus we obtain

$${}_t\breve{u}_2 = {}_t\breve{u}_1 \sqrt{\frac{3}{16\underset{u_1}{A_t}}} \cdot \qquad (11.20)$$

Whence we obtain the relation

$$16\,\underset{u_1}{A_t^3} = 3\,\underset{u_2}{A_t^2} \cdot \qquad (11.21)$$

Note that all dimensionless values like $\underset{u_2}{A_t}$ (for which the dimensions are held in the units) are mathematical-real number values. Hence, although the age of the universe (expressed in Planck event units), $\underset{u_2}{A_t}$, may take positive and negative values, the age of the universe, $\underset{u_1}{A_t}$, in atomic units must always be positive.

Thus event-scaled space can be unbounded with positive and negative time; however, our system of atomic measurement will access a superposition of the positive-time and negative-time half-spaces of hybrid space. In other words, we observe a universe with only positive time.

Now, for any practical (cosmologically short) time interval at the current epoch,

$$\underset{u_2}{\Delta t} = 2\sqrt{3}\,\underset{u_1}{A_t^{1/2}}\,\underset{u_1}{\Delta t} \cdot \qquad (11.22)$$

Image a clock ticking at a regular rate in Planck units. As the universe ages, a detector using atomic units will hear the tick rate getting faster ie time speeds up. Using the presumed constancy of c and $\hbar$, we deduce that the corresponding relationships for distance and mass values are

$$\underset{u_2}{\Delta x} = 2\sqrt{3}\,\underset{u_1}{A_t^{1/2}}\,\underset{u_1}{\Delta x} \qquad (11.23)$$

$$\underset{u_2}{m} = \tfrac{1}{4}\sqrt{3}\,\underset{u_1}{A_t^{-1/2}}\,\underset{u_1}{m} \cdot \qquad (11.24)$$

11.3.2 Large-Number Hypothesis

Dirac was the first person to suggest a connection between the fundamental mass scale and the gravitational strength in the form of an order-of-magnitude agreement

$$\frac{\hbar c}{G m_e^2} \approx \frac{A_t m_e c}{\hbar} \approx 10^{40 \pm 2} . \qquad (11.25)$$

Here m_e is the mass of an electron. Later, Robert Henry Dicke[6] argued that this order of magnitude agreement was *not* necessarily a surprising coincidence and that the order-of-magnitude agreement could be interpreted in terms of the minimum likely time required for the universe to develop and evolve a carbon-based physicist. Nevertheless Dirac's hypothesized connection remained quite a reasonable way to explain from where nature could get such exceedingly large dimensionless numbers as $\hbar c / G m_e^2$. Thus there have been numerous investigations of Dirac's large-number hypothesis using analyses of experimental cosmological data. Some significant rebuttals of Dirac-style time-dependent constants of nature are summarized by Sir Martin Rees[7].

In general, the impact of hypothesized time-dependent units is uncertain. For instance, let us consider the radius and period of a gravitational orbit as observed in atomic units. Binary pulsars can provide a very sensitive test of orbital period changes. However, even Dirac[8] entertained scenarios ranging from a contraction of the orbital radius ($\propto A_{t_{u_1}}^{-1}$) to an expansion of the orbital radius ($\propto A_{t_{u_1}}$). Both of these extreme behaviours have been shown to conflict with observation. Their derivations rely upon an Einstein-unit system in which both G and m_u are deemed time invariant. However, according to eventual reality, the quantum of information naturally depends upon the age of the universe in systems in which G is constant. Hence this Einstein-unit system is not an acceptable construction!

Eventual reality's description of nature provides four natural systems of units. We cannot categorically identify which (if any) system should display stable orbits prior to a fuller analysis of gravity, which is also likely to incorporate cosmological repulsion. However, Newton's theory of gravitation is quite adequate for most experiments when repulsion is absent. Unfortunately Newton's theory of gravity was not intended to be applied to systems with variable mass. Nevertheless, in the Newtonian limit, we might suppose orbital radii to be stable (neither spiralling in nor spiralling out) in a unit system for which $G m_u$ is time invariant[9]. Such a unit system (system 3) will have time-varying fundamental-particle masses.

Table 11.2

Unit dimensions for two further natural unit systems of interest.

Units of	System 3: Gravitational	System 4: Universal
Mass	$\frac{\hbar c}{Gm_u}$	$\frac{\hbar H}{c^2}$
Length	$\frac{Gm_u}{c^2}$	cH^{-1}
Time	$\frac{Gm_u}{c^3}$	H^{-1}
Charge	$\sqrt{36\pi\alpha_{1/3}\varepsilon_0\hbar c}$	$\sqrt{36\pi\alpha_{1/3}\varepsilon_0\hbar c}$

A gravitational unit system (system 3) in which Gm_u is invariant is shown in table 11.2. In this system, we observe that

$$ {}_G\breve{u}_3 = \frac{G^2 m_u^2}{\hbar c}. \quad (11.26) $$

Hence

$$ \underset{u_3}{G}\underset{u_3}{m} = 1. \quad (11.27) $$

Now, we may convert from gravitational units (system 3) to atomic units (system 1) Using

$$ {}_l\breve{u}_3 = \frac{3}{16}\,{}_l\breve{u}_1^2 H = \frac{\underset{u_1}{A_l}\,{}_l\breve{u}_1}{\underset{u_3}{A_l}}. \quad (11.28) $$

Hence we obtain

$$ \underset{u_3}{A_l} = \frac{16}{3}\underset{u_1}{A_l^2} \quad (11.29) $$

$$ \underset{u_3}{m} = \frac{3}{32}\underset{u_1}{A_l^{-1}}\underset{u_1}{m} \quad (11.30) $$

$$ \underset{u_3}{\Delta t} = \frac{32}{3}\underset{u_1}{A_l}\underset{u_1}{\Delta t} \quad (11.31) $$

$$ \underset{u_3}{\Delta x} = \frac{32}{3}\underset{u_1}{A_l}\underset{u_1}{\Delta x}\cdot \quad (11.32) $$

In the absence of cosmological repulsion, we anticipate that an orbit will be stable in gravitational units. Thus, in atomic units, orbits are expected to decay and slowly spiral in towards the orbital focus, the mean orbital radius decreasing as the inverse of the age of the universe.

$$r_{u_1} = \frac{3\, r_{u_3}}{32 A_{t_{u_1}}} \qquad (11.33)$$

Note that dismissals of time varying constants often rely upon an expected weakening of G causing an orbit to spiral outwards. This expectation of an outward spiral is a natural presumption but the expectation is not supported by this analysis.

Now, this estimated decay rate is too large and not supportable. However, if cosmological repulsion uniformly stretches space, then it is likely that the expansion of space will precisely cancel this orbital shrinkage leaving gravitational orbits unexpectedly stable when measured in atomic units!

11.4 Particle Charge and Colour

Eventual reality provides a physical picture of nature's elementary-particle quantum numbers. These quantum numbers reflect the observable character of the aether. The symmetries of the aether mirror the symmetries of the ground-level coordinate space; there are two sets of four branches. In each set of four branches, one branch is the 'top' scalar branch and the other branches are the quaternion-vector branches. The quaternion-vector branches will be identified both with electrical charge and with hadron colour.

11.4.1 Electrical Charge

All quaternion-vector aethereal branches are equivalent. Primarily we are not interested in which particular quaternion-vector branches hold the pertinent information; we are just interested in how many quaternion-vector branches are involved. The number (an integer from 0 to 3) of quaternion-vector branches upon which particle information resides will be called the particle's electrical charge. The charge is an integer (from an inspection of the particle table 11.3, we see that this charge is to be identified with one-third of the positron charge). Note that, as expected, the coupling interaction of the particle's information with the environment increases with the number of aethereal branches involved in holding the information. Furthermore, a charge conjugation sign $(\pm)_C$ is attached to the numerical value to allow positive and negative electrical charges.

This association of electrical charge with quaternion-vector branches provides an explanation for why the elementary fermions fit the well-known pattern of electrical-charges shown in table 11.3. Likewise, the limited range of values (from 0 to 3) for all elemental electrical charges is explained.

Table 11.3

The periodic table of elementary fermions is an arrangement of particles identified by their electrical charge and generation. Charge in measured in units of one-third the positron charge. A bar is used to denote an antiparticle.

		Electric charge (units e/3)			
		0	**1**	**2**	**3**
Generation	**0**	ν_e	$\bar{d}$	u	$\bar{e}$
	1	ν_μ	$\bar{s}$	c	$\bar{\mu}$
	2	ν_τ	$\bar{b}$	t	$\bar{\tau}$

11.4.2 Hadron Colour

Many of the electrically charged particle states just described are degenerate. We shall let hadron-colour label this degeneracy. Thus the quaternion-vector aethereal branches are individually (and arbitrarily) assigned hadron colours: negative red, negative green and negative blue. A particle with information written on only one hadron-colour branch takes the hadron colour of the informed branch. The negative hadron colour indicates that such a particle belongs to the (single electrical charge) anti-down quark family. A particle with information on two quaternion-vector branches takes the colour specified by the product of the informed-branch colours ie the negative of the colour of the remaining uninformed quaternion-vector branch. This particle belongs to the (double electrical charge) up-quark family. For these two families, there is a hadron-colour degeneracy of three. However, a particle with its information inscribed on all three quaternion-vector branches has a hadron colour of negative white colour. This particle has no hadron-colour degeneracy and it belongs to the anti-positron (electron) family group.

Thus the symmetries and the degeneracies of the hadron-colour property of elementary particles are readily explained. A charge conjugation sign $(\pm)_C$ is associated with the hadron-colour charge.

11.4.3 Generations

Information held by each aethereal branch of an upper phase-space is written in up to three quaternion-vector directions. All vector directions are equivalent. Thus it is the number of directional dimensions involved that is physically relevant and provides a distinguishing feature that may be used to identify the appropriate quantum number. This number of quaternion-vector dimensions is referred to as the generation index.

Hence, for the first generation (generation index equal to zero), the information is displayed radially in all three dimensions (ie spherically). For the second generation, the information is displayed in a bimodal form (hourglass or dumbbell shape), etc. Now, potentially we could have a degeneracy of three for particles of the second and third generations. However, unlike the branch-labelled hadron-colour there is no physical discriminatory label: there are no special symmetry-assigned (quaternion-vector) directions[10]. Consequentially particle members of the second and third generations are a simple superposition of these possibilities. This superposition provides a multiplicity factor of three that appears in the mass formula (to be discussed) for these generations.

The requirement that information be written in a phase space determines that there can be no particles when no upper aethereal directions are involved. In this way, the hierarchy of fermion generations is defined. There are three independent quaternion-vector directions and hence just three fermion generations, each generation with a set of distinguishable particles possessing hadron-colour and electrical charge.

11.5 Mass Terms

One cannot observe isolated quarks. This undesirable fact leads to complications when considering quark masses. Thus experimentally we only have direct access to dynamical quark masses. Dynamical masses are obtained by parameterizing the exchange of energy and momentum in collisions between an observable particle and a seemingly loosely bound quark located within a hadron. These dynamical masses (or effective masses) approximately describe the shape of the quark's band structure; dynamical masses can differ greatly from Newton's idea of mass, the quantity of substance. However, a different approximation gives us a possible way to estimate the quantity of substance.

If the surrounding sea of virtual quarks does effectively spread the hadron-colour of a quark over a significant volume it could turn this bloated troupe of particles into a more weakly interacting quasi-particle. Then we could estimate the mass of these quasi-particles from their component contribution to the measurable masses of known elementary particles. These presumed quasi-particle masses are known as constituent masses.

In light of these uncertainties the following hand waving arguments will mainly relate to the charged leptons.

11.5.1 Self-Energy Interactions

The analysis of the Dirac equation has shown that the rest energy of a particle is made from two terms, m_1 and m_2. One term holds the phase-space information requirement; the other term represents an interaction self-energy contribution.

Without self-energy contributions, the rest masses of elementary particles would show quantized behaviour. Consider firstly electromagnetic contributions to the self-energy. The dimensionless electromagnetic strength should be given by the square of the unit charge in some natural units, eg

$$\alpha_{1/3} = \frac{(e/3)^2}{\varepsilon_0 hc} \qquad (11.34)$$

Thus we gauge the electromagnetic self-energy contribution to mass could be

$$m_2 = \alpha_{1/3} n^2_{charge} m_{\shortparallel} \qquad (11.35)$$

where n_{charge} is the integer electrical charge. This estimate can also be obtained from the quantum uncertainty principle. Identifiable elements with energy $m_{\shortparallel}c^2$ cannot be squeezed into a distance closer than $\hbar/(2m_{\shortparallel}c)$. Then, the electrostatic potential energy of the stable ground-state configuration is taken to be

$$P.E. = \frac{n^2_{\text{charge}}(e/3)^2\, 2m_{\shortparallel}c}{4\pi\varepsilon_0\hbar} = \alpha_{1/3} n^2_{\text{charge}} m_{\shortparallel} c^2 . \qquad (11.36)$$

Now, quarks fail to span[11] all higher vector branches. The empty branches can be chosen in $\binom{3}{n_{charge}}$ ways. For quarks of all flavours, the self-interaction of these superposed arrangements could be deemed to contribute approximately

$$m_2 = 9m_{\shortparallel} \qquad (11.37)$$

to the rest mass.

Now, for the current era and to within an accuracy of about 10%, the Hubble 'constant' is measured to be 72 km/s/Mpc[12]. This Hubble value corresponds to an age of the universe equal to 13.6 giga-years and a mass unit, m_u, equal to 35 MeV.

Table 11.4

Numerical values of estimated contributions to the self-energy terms for elementary fermions. Electric charge is measured in units of one third the positron charge.

Electric charge	0	1	2	3
Electromagnetic self-energy	0	57 keV	227 keV	511 keV
Hadron-colour self-energy	0	315 MeV	315 MeV	0

Computed values for the estimated self-energy contributions are given in table 11.4.

11.5.2 Leptons

Lepton information extends over each of a set of four upper aethereal branches. Indeed, it is a reflection of the lack of constraints (premise 1) that all possible allocations of directions (limited by the number of available dimensions) are superimposed upon the structure identifying the particle.

The first directional dimension only signifies existence. A second direction signifies one count of information (ie one unit of mass). A third direction gives two counts of information. Thus if each identifiable branch has information written in a number, $i_d + 1$, of higher-level directions, then a set of branches holds i_d^4 units of information. However, each higher generation has subcomponents with lower symmetries. Hence, the total count of information is made from a summation.

Thus (for generation index n_{gen}) the total count of unit information across the vector branches is

$$N = 3 \sum_{i_d=0..n_{gen}} i_d^4 . \qquad (11.38)$$

Consequently we deduce that the information count, N, takes values 0, 3 and 51 respectively for the three generations of leptons.

Now, (using the calculated mass unit, m_u, of 35 MeV) the phase-space mass, m_1, takes values equal to 0, 105 and 1785 MeV for the three generations of triply charged fermions. Recall that an estimate of the rest masses of these particles is given by the sum of this phase-space mass and the self-energy.

Hence, these plausible estimates are in reasonable agreement with experimental values. However, these arguments were fabricated and factors were deliberately massaged in the full knowledge of the experimental masses.

Notes

[1] Here the brackets are used to denote a function.

[2] Conventionally the phase-space cell has minimum size, $(\Delta x \Delta p_x)_{min} = ½\hbar$. The corresponding dimensionless expression is $(H\Delta x / c)(\Delta p_x / m_1 c) = h_0$.

[3] When Đ is constant, the mass unit, $m_{II}c^2 = \hbar H / 2Đ^{2/3}$, varies inversely with the age of the universe.

[4] This basic argument seems to have been first formulated by Sivaram (1982).

[5] $\alpha_{1/3}$ is proportional to the fine structure constant. $\alpha_{1/3} = (e/3)^2 / \varepsilon_0 hc$. We shall show that eventual reality regards $e/3$ as the fundamental unit of electrical charge.

[6] Dicke (1961) discusses how the first generation stars make higher elements (beyond Helium) in the periodic table. This material is dispersed when the stars die. Then second generation stars like the Sun can form with rocky planets hosting the evolution of carbon-based life forms. Such life forms are needed to note the apparent similarity of these large numbers.

[7] Rees (1997) pp236-239 lists many tests that have failed to observe any variation in the relative strength of gravity. Currently there is tentative evidence only for a very slight variation in the fine structure constant.

[8] Unfortunately with only coincidences of large numbers to suggest associations of 'constants' with time, one cannot unambiguously decide on the form of any temporal dependence. If the number of nucleons (in the observable universe) increases as the square of time then matter is apparently being created. Depending upon where and how matter is created, Dirac (1973) has shown that gravitational orbits may spiral in or spiral out.

[9] In Newtonian analyses, the orbital radius and orbital velocity are stable when both Gm_{II} and c are invariant.

[10] The symmetries of the branches are identified with spatial parity, time reversal and charge conjugation. However, the spatial directions are all equivalent.

[11] This incompleteness is responsible for their hadron-colour assignment.

[12] This value has been taken from Bartusiak (2004).

Part 2

Chapter 12 : Beneath Quantum Reality

The theory of eventual reality does more than simply provide an explanation for quantum reality. Eventual reality can provide answers for all the great mysteries. We shall start the articulation of meta-physical explanations by considering how we can describe the aethereal reality that lies behind the material quantum world[1]. This exposition will expose an explanation of mind, mathematics and logic.

12.1 Launching Point

The meta-physical arena is located beyond science. Explicitly we shall restrict the term *science* to that set of scholarly disciplines that describe and explain the *quantifiable* aspects of the universe.

Recall that such science outwardly gives the impressions that (1) almost everything is explainable and (2) almost everything falls within the applicable scope of this quantitative science. However, our experiential reality is highly qualitative. Thus direct experience is likely to be beyond the explanatory powers of any science based upon the quantitative. For instance, emotions, desires, concepts and ideas are substantially qualitative entities. These entities are products of the working of the mind. However, the reader should not get the impression that the conceptual rift between the quantitative and the qualitative is equivalent to the mind-body problem. Experience reveals many other natural qualitative properties that cannot be fully made quantitative eg shape and structure (examples of a representation and a classification).

Other detailed products of the mind such as a story, a discussion, a new theory, leadership, a design, a plan or a game are also principally qualitative entities although these complex products may have many quantifiable qualifiers and even quantitative implications. Essentially quantitative features of reality (describable by science) are just those residual elements of a 'dry' reality that has been drained of its intangibles, its implicit meritorious values.

For the last quarter of a century, some scientists have been proclaiming that science is close to discovering[2] a theory of everything (that is quantitative) ie a theory upon which the physical sciences can be founded. Eventual reality is quite unlike the previous contenders and pretenders for this honour. Eventual reality is anchored to a qualitative reality; the statistics of this qualitative reality are claimed to generate a rich and quantitative scientific realm. One aim of this chapter is to present other implications of this qualitative foundation.

Note that scientific theories, themselves, are substantially qualitative (ie descriptive) in their nature. Similarly arguments about the existence and development of science are non-quantitative. The scientific method can involve the use of synthesis, analysis, analogues, prediction, deduction, conjecture and reason. However, many scientists also talk about the beauty of a fundamental solution and the joy of the construction of an analysis. Clearly science is impeded from explaining its own processes. These processes of science must reside outside of the predictive and explanatory scope of current science since science's paradigms and beliefs are inherently qualitative. Therefore it is likely that only a qualitative theory could be able to account for science's existence and elucidate why the world is describable by a science. Potentially such a theory can explain much more. Thus another aim of this chapter is discuss how such qualitatively based theories can necessarily generate a polymorphic universe of space, matter, science, mathematics, logic and mental processes.

A multi-branched structure for space is a seminal prediction made by eventual reality; however, taking a decidedly unscientific posture, eventual reality does not allow the quantitative features on the branches to be independently inspected by conventional experiment. Clearly the theory of eventual reality does not accept the quantum belief that only measurement results are to be regarded as real. Furthermore, with eventual reality's broad scope (covering both the qualitative and the quantitative) and with some quantitative features that are scientifically undetectable, much of eventual reality has to be attributed to the broader discipline of natural philosophy rather than be attributed to the narrower disciplines of the physical sciences. However, it should be noted that it would have been difficult for a philosopher (as opposed to a scientist) to have developed this theory.

Now, beneath the quantitative framework of space, the universe is an ensemble of referential events. On a large scale, we can employ forms of analytical smudging to generate analytical quantitative scientific descriptions. These quantitative features such as the approximate location of an event, its quantitative relationship to its near neighbours and the properties of a region of space are determined collectively by all the (qualitative) relationships within the entire environment of an event.

The coherent stubble-like structure that is attached to the branches of space is referred to as the aethereal field. The aethereal field that pervades the branches of space-time maintains an implicitly qualitative value. The quantitative details are generated statistically and hence the generated scientific description (eg quantum mechanics) is probabilistic.

12.2 Representation of the Qualitative

To further our understanding of the universe, we need an appropriate way to describe the qualitative characteristics of the aethereal field. Now, mathematics is sometimes said to be the language of science. Just as our forbears discovered the forms of mathematics and used mathematics to describe the quantitative features of our world, our societies developed human language as a tool to describe the general features (particularly the qualitative features) of our world. The power of (qualitatively based) language is revealed by the evidence that language can even be used to describe language itself.

Now, any symbolic system that is in one-to-one correspondence with a language is also a language. Thus we assert that the aethereal field *is* a library of literary works expressed in *the* universal langue. We shall use the properties of this aethereal field to define an acceptable version of this abstract universal langue. In this langue, we shall not distinguish a lexeme (word or word stem) from the lexeme's meaning.

To help us to comprehend the universal langue, we shall expose some connections and general similarities of the universal langue with familiar languages. In particular, we recognize that languages possess elements, elemental categories, structural rules and assignments. However, just as Schrödinger's quantum wave theory for the propagation of probability-amplitude waves owes little to the preceding classical wave theories, a universal langue should not be expected to closely resemble any familiar language.

The atomistic elements of language are the lexemes. The meanings of lexemes can be expressed in ordered sequences of words; as such, the lexemes and their definitions can be assembled into a dictionary. Primarily meaning is conveyed in the way these lexemes relate to one another within the entire lexicographical environment. For instance, a consistent replacement of the representation of one lexeme with another representation throughout the dictionary will not alter any meanings. Note that this self-referential generation of meaning, parallels the way that events create space. Further the secondary naming of facts, artefacts, actions and reactions in regular languages, is only linguistically important in transferring an assignment of meaning.

The aether is an extension of space-time, thus the universal lexicon includes all possibilities for the coherent assembly of events. Consequently we expect that the universal langue will share features in common with our previous quantitative analyses. Though, by necessity, the corresponding arguments will be slightly modified.

Lexicographical space does not have numeric extension; we do not need translations and rotations. Instead we have a requirement for stable digital information derived from discrete pointer links. Stability dictates stationariness; stationariness decrees group symmetry; group symmetry delivers an algebra.

We want to be able to combine ($\times$) the algebraic elements across different branches and to form course-grain averages ($+$) over volumes on the same branch. Thus we again seek an algebra with binary (ie two) operations. Most of the requirements for a division ring derive from the individual group-symmetry requirements. The extra requirement is the distributive law. The distributive law describes how we work with both these × and + operations enabling us to appropriately combine the results across branches and then take a collective-coordinate average. For instance these averages should be combinable extensively across neighbouring regions. Given that strengths of averaged digital information pass muster as mathematical-real numbers, the stationary aether is described by the space-like symmetries.

Thus we conclude that a universal langue will be representable as a hierarchy and that the langue will display repeated sets of mathematical-complex, SU(2), and quaternion, Q_8, symmetries.

We can contrast our stipulations for a representation of the universal langue with the state of human languages. Our idiosyncratic spoken languages have limited hierarchical structures and contain unnecessary ambiguities in the meanings of words. A dictionary entry that contains a multiple list of meanings for a word is the norm rather than the exception. Furthermore, our regular languages allow sentences that cannot be true:

The petunias ate my homework.	*(non-credible)*
Canberra is the capital of Turkey.	*(non-factual)*
This sentence is false.	*(illogical/undecidable).*

However, our spoken and written languages do contain a comprehensive vocabulary that should be translatable into and from the universal langue. In contrast, human-designed computer languages are more regular and consistent but have a specialized vocabulary. In particular 'Object Orientated Programming System' (OOPS) languages, with a hierarchical structure of objects belonging to generations, with property acquisition through inheritance and with unambiguous lexeme meaning, have many organizational features expected of the universal langue. Thus we shall construct a representation of the universal langue using the vocabulary derived from human language and a structure like an OOPS computer language.

So let us look at an empirical assignment of meaning to these repeated SU(2) and Q_8 symmetries by considering how these symmetries feature in known languages. Later we shall attempt an *a priori* rationalization of these assignments.

12.2.1 Lexeme Flavours

In the previous chapter, we have noted that the elementary fermions come in a range of flavours. Similarly we shall show that the primal lexemes can be attributed flavour-like properties that are associated with their quaternion symmetries.

12.2.1.1 Quaternion Symmetry

We use language within our thoughts. Thus we cannot separate our use of language from the way in which we think. Not surprisingly Carl Jung's psychological types don't only tell us about people's nature, they reveal entrenched detail about the structure of human language and thought. His psychological types are commonly encountered through the Myers-Briggs type indicators, which describe the preferred ways in which our minds form and handle impressions of the world. Implicitly we shall assume that this Jungian analysis is complete, that this Jungian analysis is relevant and also that there are no additional mental degrees of freedom that are inaccessible to the human mind ie there are no missing avenues of mental functioning that humans cannot access.

Historically there is background support for accepting the premise that representations, espousing four independent degrees of freedom, may be adequate to describe the qualitative realm. There have been well-accepted representational schemes with fewer degrees of freedom but rarely more than four. Recall that most sages of old from the time of Empedocles (who promulgated the pluralistic view that everything was composed of mixtures of Air, Earth, Fire and Water) have been content with a four-dimensional description of much of reality[3]. The physiology of the medieval world held that one's state of mental and physical health was determined by the four humours (blood, yellow bile, phlegm and black bile). Independently the seventeenth century theologian Antoine Arnauld[4] identified four principle operations of the mind. He labelled these operations as conceiving (1), judging (2), reasoning (3) and ordering (0). Note that the appended integers, which I have assigned to these mental operations, will be explained later.

Clearly any relationships (related to the psychological types) that are displayed by human language should hold also (after translation) in the universal langue. Furthermore, such a four-dimensional description of aspects of the universal langue would fit with the symmetry (Q_8) of quaternions.

Recall that the elementary fermions (see table 11.3) can be arranged in a flavour table of particles according to their electrical charge and their generation index. The number of independent degrees of freedom fixes the number of generations; however, the generation index does not correspond to a particular labelled dimension.

Something similar is true for our primal lexemes. Intrinsically aethereal quaternion-vector branches and aethereal quaternion-vector directions have the same symmetry structures. Furthermore, the most natural representation will not differentiate between different quaternion-vector directions nor differentiate between different quaternion-vector branches. Thus the flavour table for lexemes, will be concerned with the number of aethereal directions and the number of aethereal branches that are involved in the representation, not which particular quaternion-vector directions or which particular quaternion-vector branches are involved.

CHAPTER 12

Table 12.1

The table lists the four indicator labels that are applied to the Jungian psychological dimensions by a Myers-Briggs analysis of preferences. The table also provides descriptors of the related range of mental functioning. The associated numerical generation/spectrum index does not belong to the Myers-Briggs analysis and this index is explained in the text.

Generation / spectrum index			External focus	Internal focus
0 (J-P)	decide	order	'Judging'	'Perceiving'
1 (E-I)	interact	create	'Extrovert'	'Introvert'
2 (S-N)	detect	understand	'Sensing'	'Intuitive'
3 (T-F)	argue	motivate	'Thinking'	'Feeling'

Let us show how to assign the appropriate generational ordering for the linguistic associates of the psychological traits. This ordering has already been anticipated in the layout and in the embedded numbering in table 12.1. We will present four 'spectra' of what will be termed mental processes and mental notions. Each spectrum has a different flavoured character; each table is associated with a different psychological trait. We shall start this inventory with the spectrum with which many physicists should feel most comfortable. This spectrum (#2) deals with how we move from data to belief. The spectrum parallels the structure of science which is built upon observation (ie gathering information), law (ie interpreting the regularities of observations), principle (ie subsuming laws to create meaning) and paradigm (ie a framework of beliefs). This sequence of assembled stages has an unambiguous order.

This information spectrum is presented in table 12.2. This spectrum provides an ordering of mental notions: information, knowledge, meaning and belief and the spectrum provides a corresponding ordered sequence of mental processes: gathering, interpreting, understanding and exploring relationships. However, the processes in this table do not look so independent: Does not interpreting involve exploring relationships and understanding? Does not understanding involve interpreting and exploring relationship? These relationships actually reflect aspects of the underlying relational symmetry of the components of a quaternion-vector space.

Now, each line in the spectrum carries the taint of a psychological trait. Hence, the evident spectral order of the entries (eg table 12.2) allows us to sequence these psychological traits. In turn, this sequencing is used to number the spectra and number the generations. There is an equivalent spectrum for each of the other numbered psychological traits. Thus the information spectrum is spectrum 2. The other spectra are given in tables from table 12.3 to table 12.5.

Table 12.2

The second spectrum (S-N) parallels the hierarchical development of scientific understanding. This spectrum spans the real information highway that marks how we build and develop a realistic understanding of the world with its accompanying belief structure.

Generation no.	Process	Associated indicator	Notion	Building Spectrum
0	Gathering	J-P	Information	Interpret (information) = Knowledge
1	Interpreting	E-I	Knowledge	Understand (knowledge) = Meaning
2	Understanding	S-N	Meaning	Explore relationships (meanings) = Belief
3	Exploring relationships	T-F	Belief	Accept (belief) =Faith

Table 12.3

Spectrum 0 (J-P) builds the processes of collection and categorization.

Indicator no.	Process	Indicator	Notion	Building Spectrum
0	Get	J-P	Observation	Analyse (observation) =regularity
1	Analysing	E-I	Regularity	Review (regularity) =classification
2	Reviewing	S-N	Classification	Explore options (classification) =choice
3	Exploring options	T-F	Choice	Accept (choice) =uncertainty

Table 12.4

The first spectrum (E-I) is associated with the activities of communication, learning and creativity.

Indicator no.	Process	Indicator	Notion	Building Spectrum
0	Retrieve	J-P	Occurrence	Contemplate (occurrence) =idea
1	Contemplate	E-I	Idea	Understand (idea) =viewpoint
2	Understand	S-N	Viewpoint	Exploring (viewpoint) =prediction
3	Explore	T-F	Prediction	Acquiescing (prediction) =enlightenment

Table 12.5

Spectrum 3 (T-F) handles how we feel about things.

Indicator no.	Process	Indicator	Notion	Building Spectrum
0	Receive	J-P	Sensation	Dwell on (sensation) =desire
1	Dwell on	E-I	Desire	Conceptualize (desire) =enthusiasm
2	Conceptualize	S-N	Enthusiasm	Tempt (enthusiasm) =commitment
3	Tempt	T-F	Commitment	Surrender (commitment) =peace

The set of processes for these spectra are similar and on stripping away stray connotations the core functioning of all spectra can be seen to be approximately represented by the common set of primal instructions: access, compare, give significance and test.

In building the spectra, we observed that the spectral processes could generate the spectral notions. Each spectrum starts with datum that is flavoured with its corresponding psychological trait. Thus we represent the spectral-start notions, which are the notions that relate to the 'external' world, in the form

$$\begin{bmatrix} \text{observation} \\ \text{fact} \\ \text{information} \\ \text{sensation} \end{bmatrix} = \text{access} \begin{bmatrix} \text{datum}_0 \\ \text{datum}_1 \\ \text{datum}_2 \\ \text{datum}_3 \end{bmatrix} = \text{access}(\text{input data}) \cdot \qquad (12.1)$$

Consequently the full set of primal notion is given in table 12.6.

Table 12.6

The 'periodic' flavour table of primal notions associated with the directions and branches of the qualitative aethereal field.

	J-P (0)	E-I (1)	S-N (2)	T-F (3)
J-P (0)	Observation	Occurrence	Information	Sensation
E-I (1)	Regularity	Idea	Knowledge	Desire
S-N (2)	Classification	Viewpoint	Meaning	Enthusiasm
T-F (3)	Choice	Prediction	Belief	Commitment

Language, as we know it, is imbued with the primal sequence of processes 'access, compare, give significance/comprehend, test'. Such a primal sequence is in agreement with a recognizable universe possessing free will, predictive science, comprehensive beliefs and emotional commitments. Thus we could employ empiricism to argue for this assignment of these notions and processes with qualitative directional features on the spatial branches. However, this does pose the question: 'Could language exist without this structure?'

Instead we argue the following steps: 1) the universe must display the quaternion-style structure; 2) language (or indeed any imaginable whatsoever) may only exist within a universe; 3) within the scope of their vocabularies, all languages can be translated into each other; then 4) every comprehensive language (and in particular the universal langue) must possess a primal quaternion-style symmetry.

Simple languages (eg body language) exist to provide a means of instinctive communication. More complex languages exist to effect an active engagement with an environment. To actively communicate, language is formed into a descriptive interaction. A proto-langue, expressible in the first level of the aethereal hierarchy, obeys the rules of group symmetries. We shall discuss later how the mathematical signs of designator (eg $+_p, -_p$) denote opposite values and hence how these signs give rise to Boolean logic. Such logic enables relative communication of qualitative values. This Boolean logic and a number system (eg coordinate) comprise the foundation of the sub-langue of mathematics.

Next, we shall consider the SU(2) group that must accompany the quaternion (Q_8) group. This SU(2) group will also be accompanied by a mathematical sign.

12.2.2 Complex Symmetry

All known languages have a grammatical structure. In many familiar languages, grammar can becomes quite technical with many parts-of-speech (such as verb, adverb, noun, pronoun, adjective, preposition, conjunction, interjection, etc.). However, we shall differentiate the contextually non-local referential pronouns, determiners, prepositions, conjunctions and sentence connectors that either link or refer to other sections of text from the core contextually local language elements (verb, adverb, noun, adjective and interjection).

We have observed in coordinate space that the components of mathematical-complex coordinates formed conjugate variables. These conjugate variables were observed to have several linkages. For instance, the mathematical-real component describes the capacity of the mathematical-imaginary component to change (ie the real component describes the activity of the imaginary component).

Hence, these observations lead to the natural suggestion that we should associate the mathematical-real aethereal component with process (ie the verbs) and associate the mathematical-imaginary aethereal component with its corresponding notion (ie the nouns). The observations reveal that the required mathematical-complex SU(2) symmetry is associated with simple grammar.

The necessity of using mathematical-complex expressions, requires that we write

$$\begin{bmatrix} \hat{r}\,\text{access} + \hat{i}\,\text{observation} \\ \hat{r}\,\text{access} + \hat{i}\,\text{fact} \\ \hat{r}\,\text{access} + \hat{i}\,\text{information} \\ \hat{r}\,\text{access} + \hat{i}\,\text{sensation} \end{bmatrix} = \text{access} \begin{bmatrix} \hat{r}\,\text{do}_0 + \hat{i}\,\text{datum}_0 \\ \hat{r}\,\text{do}_1 + \hat{i}\,\text{datum}_1 \\ \hat{r}\,\text{do}_2 + \hat{i}\,\text{datum}_2 \\ \hat{r}\,\text{do}_3 + \hat{i}\,\text{datum}_3 \end{bmatrix} = \text{access}\left(\hat{r}\,\text{do} + \hat{i}\,\text{data}\right) \qquad (12.2)$$

$$\begin{bmatrix} \hat{r}\text{ compare} + \hat{i}\text{ regularity} \\ \hat{r}\text{ compare} + \hat{i}\text{ idea} \\ \hat{r}\text{ compare} + \hat{i}\text{ knowledge} \\ \hat{r}\text{ compare} + \hat{i}\text{ desire} \end{bmatrix} = \text{compare}\begin{bmatrix} \hat{r}\text{ do}_0 + \hat{i}\text{ datum}_0 \\ \hat{r}\text{ do}_1 + \hat{i}\text{ datum}_1 \\ \hat{r}\text{ do}_2 + \hat{i}\text{ datum}_2 \\ \hat{r}\text{ do}_3 + \hat{i}\text{ datum}_3 \end{bmatrix} \qquad (12.3)$$

$$\begin{bmatrix} \hat{r}\text{ understand} + \hat{i}\text{ classification} \\ \hat{r}\text{ understand} + \hat{i}\text{ viewpoint} \\ \hat{r}\text{ understand} + \hat{i}\text{ meaning} \\ \hat{r}\text{ understand} + \hat{i}\text{ enthusiasm} \end{bmatrix} = \text{understand}\begin{bmatrix} \hat{r}\text{ do}_0 + \hat{i}\text{ datum}_0 \\ \hat{r}\text{ do}_1 + \hat{i}\text{ datum}_1 \\ \hat{r}\text{ do}_2 + \hat{i}\text{ datum}_2 \\ \hat{r}\text{ do}_3 + \hat{i}\text{ datum}_3 \end{bmatrix} \qquad (12.4)$$

$$\begin{bmatrix} \hat{r}\text{ test} + \hat{i}\text{ choice} \\ \hat{r}\text{ test} + \hat{i}\text{ prediction} \\ \hat{r}\text{ test} + \hat{i}\text{ belief} \\ \hat{r}\text{ test} + \hat{i}\text{ commitment} \end{bmatrix} = \text{test}\begin{bmatrix} \hat{r}\text{ do}_0 + \hat{i}\text{ datum}_0 \\ \hat{r}\text{ do}_1 + \hat{i}\text{ datum}_1 \\ \hat{r}\text{ do}_2 + \hat{i}\text{ datum}_2 \\ \hat{r}\text{ do}_3 + \hat{i}\text{ datum}_3 \end{bmatrix}. \qquad (12.5)$$

The grammar is not entirely that of nouns and verbs. The scalar term is of the form of a mathematical-complex pair: (verb, noun); however, the grammatical terms on the vector branches may be more appropriately viewed as qualifiers of the scalar-branch term ie the vector-branch terms are $\hat{r}$ adverb + $\hat{i}$ adjective. However, such subtle details will be ignored in this overview of the universal langue.

Note that we have displaced our prior conceptions of a material universe by a universe in which the currency of the aether is language. From this perspective, we shall consider whether the spectra of mental activities and mental notions can be seen to have a natural basis.

Thus let us try to make sense of the approximate identification of the elemental set of processes (access, compare, understand, test) by considering what meaning might reasonably be attached to the action of pointer links in a coherently networked environment. A single link is a pointer to something and thus the link tells us where to access information for any activity. (Since the something, which a pointer-link identifies, has its own links, we may regard this something as a compound-data set rather than simply another pointer). Nominally we use a single link to signify the quaternion-scalar direction since there is no relative information to which we may relate the link. Clearly it is very reasonable to align this single-dimension structure with a referencing activity such as *access*. However, double links provide the opportunity for comparisons of the referenced data. Indeed, since a double reference link is not the same as two separate single links, a double link could only correspond to comparison. Retaining the convention employed for a single-link, the quaternion-scalar direction indicates the item in focus. As with elementary particles, a single vector-direction structure does not preference any particular spatial vector direction, rather the information structure points isotropically. Thus the pointer makes a contrast between a data set and an undifferentiated surrounding value. Hence, it is reasonable to assign these single-vector-direction aethereal structures to the activity of *comparison* or qualitative assessment. Now, a dual vector-direction structure (triple link) must build upon a comparison. Thus a dual vector-direction structure can differentiate comparisons. An aethereal structure with this feature focuses the comparisons in the surrounding constructed knowledge. This set of relationships is attributed to the activity of relating one structure to the specified correlations of the embedded environment, an activity that we shall abbreviate simply as to *give significance*. Finally the most complex of aethereal structures is the superposition of the most structured of possible realities. This superposition is sampled, compared and tested by whatever challenges the environment holds. Hence, we reasonably called this process to *test*.

Although instructive, this line of reasoning cannot provide *definitions* of the fundamental lexemes. Quite simply the meanings of individual lexemes are not deductively assignable in such a fashion. Such rationality is most suited to factual quantitative and logical environments. Indeed the words we have used are really too high-level for the tasks that they need to describe. Words and fundamental reality are inextricably ensconced in qualitative media. The individual attribution of a lexeme is derived from the totality of the relations of the linked medium. Thus meanings are more sophisticated than logical facts – this is the significance of the above argument. The basic processes (access, compare, give significance and test) form a natural sequence that is ordered by increasing referential sophistication. The difference between the words is only a difference between their burdens of relational associations. In normal language, the connotations that we place upon the words of this sequence disguise these words' core role in the formation of all words possessing compound sense.

12.3 Logic

Logic is an essential part of any linguistic assignment to the aether. Logic is manifest in language through the art of rational thinking, and logic appears explicitly in all languages. The primal lexemes are associated with designators (quaternion and mathematical-complex), leaving the mathematical signs to be associated with logical assertions. Note that the mathematical sign symbolizes a reversal of direction of the associated designator/lexeme. Furthermore, we have the sign-product rule

$$(+)(+) = (-)(-) = (+). \qquad (12.6)$$

Thus true (+) and false (-) are absolutely identified; true and false are not interchangeable or arbitrary. However, with attached flavour they can become transformed into good and bad, new and old, right and wrong, etc.

Now, symmetry requires that we should use full SU(2) grammatical expressions. An example of such a full expression would be

$$(\pm)_P\left(\hat{\mathrm{r}}\,[\mathrm{test}] + (\pm)_T\,\hat{\mathrm{i}}\begin{bmatrix}\text{choice}\\ \text{prediction}\\ \text{belief}\\ \text{commitment}\end{bmatrix}\right) \qquad (12.7)$$

which contains two non-equivalent signs. The mathematical signs independently act on the process and the notion.

Thus this SU(2) symmetry reveals that logical expressions may evaluate to one of the four outcomes like

$$\hat{\mathrm{r}}\left\{{}^{test}\!\big/_{untest}\right\} + \hat{\mathrm{i}}\left\{{}^{true}\!\big/_{untrue}\right\} \qquad (12.8)$$

Here 'untest' (the negative of test) is used to signify a reversal of a test process. This negation should not be confused with the idea of an opposite, which more often signifies a reversal of the result of the process.

Now, we have used classical logic in the presentation of this work. However, the natural logic that we deduce from eventual reality is different from the presented classical logic. Hence, we clearly see that we are not trivially getting out of eventual reality the logical rules that our arguments have put into the theory.

Historically[5] there are three laws of classical logic:

1) Law of non-contradiction: a statement cannot be both true and false.
2) Law of identity: if A is identical to B, then whatever is true for A is also true for B.
3) Law of excluded middle or law of bivalence: every statement must be either true or false.

The law of the excluded middle (*tertium non datur*) is problematic. Our experiences in a world that approximately adheres to the principles of classical mechanics lead us to believe in the certainty of either the validity or falsity of all statements. Hence, like classical mechanics, classical logic has a problem with free will. For instance, classical logic has a difficulty with a statement like *'At the next election, the government will be defeated'.* Intuitively we know that the validity of this statement is conditional, it is neither true nor false. Classical logic maintains that this statement does have a logical value even though we currently do not know what that value is. Classical logic believes in fate rather than free will.

Furthermore, the law of bivalence does not accept '*This statement is wrong*' as a valid statement. The universe is more lenient with the self-referential. Without calamity, we can contemplate our own thought processes. Nature is similarly self-referential, implicit equations in physics often provide self-consistent sets of relations which may be manipulated and computationally solved without contradiction.

Commonly to overcome the logical difficulty with '*This statement is wrong*', a sentence that refers to any statement may be called a meta-statement. However, this approach displaces the problem; the approach does not resolve the issue. Thus a sentence that refers to a meta-statement (eg '*This meta-statement is wrong*') needs to be called a meta-meta-statement, etc. Apparently a resolution could be afforded by allowing a sentence that refers to a meta-statement to be called a statement. This bifurcated resolution of the logical difficulties would introduce two independent sets of true and false (one set for a statement and one set for a meta-statement); these two sets parallel the four logical outcomes displayed by the symmetry of eventual reality.

Another recognized approach to overcoming these dilemmas is to remove the law of excluded middle from the premises of logic. One system of logic (based only upon the first two classical laws) is referred to as intuitionistic logic[6] and its corresponding mathematical field is known as intuitionism. The focus of intuitionistic logic is on the verbal proof of a proposition. Thus in intuitionism, the absurdity of a proposition equating to $\hat{r}\{test\} + \hat{i}\{true\}$ does not imply that the particular proposition must evaluate as $\hat{r}\{test\} + \hat{i}\{untrue\}$. Hence, conclusions provided by the procedure known as *reductio ad absurdum* will often become inconclusive. Examples of such proofs include St Anselm's ontological argument for the existence of God and Russell's paradox.

Now, eventual reality implies the need for a clear differentiation between verbs and nouns and a clear differentiation between their associated signs. Thus the double Boolean logic of eventual reality may be viewed as a particular version of intuitionistic logic.

This logical option, of undecidedness for a statement, is the inevitable conclusion of eventual reality, of Gödel's theorem and of the quantum fable of Schrödinger's cat. Within the theory of eventual reality, undecidedness has been given a natural and rational justification.

12.4 Mathematics

The field of logic is an important part of the discipline of mathematics. However, mathematics has additional concepts and hence mathematics can claim to be a larger field than logic. Consequently mathematics cannot be founded upon logic alone. Now, the foundation storey of eventual reality contains an infinite set of countable events. These events provide a direct universal rational for mathematical integers; this rational for the event's abstract existence is independent of the system of logic. The integers will define the mathematical operations of multiplication and addition. In turn, multiplication and addition (plus the system of integers) define the system of real numbers. From this structure of logic and the mathematical-real numbers, diverse fields readily stack up into the familiar intellectual achievement known as of mathematics.

Now, we need to clarify the justification for logic. It is within the context of the justification of mathematics that we have justified intuitionistic logic. Whilst it may seem that we are using premises, which we are attempting to prove; we are actually employing the *by inspection* method. Thus the core logical relations come simply from the algebra of the arithmetic signs. A requirement for arithmetic signs becomes inescapable when we consider inverse operations for multiplication and addition. Mathematics arises from the ordinal countability of sets of events and the coherent spatial-like structures dominating the network of events. The self-consistency of the layered justifications demonstrates the natural symmetries to be exposed in the coherent universe. We have already argued for these symmetries.

12.5 Mind

The universal langue and matter are intimately mingled. The universe bubbles with opportunity for linguistic expressions that are formed from relational associations of appropriate events. The pattern of these associations defines the freedom of expression; hence the pattern defines the probability of achieving any particular form. The multiplicity of probabilities over branches defines the quantum state of matter. In the absence of imposed control/will, predictable quantum-mechanical chaos reigns over this world. Thus the feverish hum of over-expressed language begets matter. In turn, local matter influences the local field of expressed gossip. For instance, measurement has a dramatic affect upon the quantum wavefunction (ie freedom of expression); measurement causes messages to be substantially rewritten across all branches. Similarly, as matter spreads and propagates throughout space, the underlying format of ideas spreads with the associated matter in time. However, ideas (however they arise) are compelled to disperse and interact with surrounding competing ideas.

Sometimes, a gossip-matter loop can be organized into a self-sustaining feedback loop. A feedback loop offers a capability for the locally contained control of both matter and linguistic ideas. Note that linguistic ideas formed under the apparent control of matter are *thought*. In turn, the thoughts of this feedback loop have the capacity to manipulate local matter. Temporal repetition/preservation of thought can also benefit from the existence of a memory capability. Robustness requires that memories should be stored entirely in classically sustained ways.

Like an acorn, these feedback loops can grow in sophistication. As the loops acquire knowledge and experience, their structure of linkages build. In a surrounding sea of noise, the growth of disorder is inevitable. This sea of noise is generated by random ideas propagating from elsewhere. However, enveloping systems that degrade externally sourced energy can protect the decay of a sophisticated feedback loop. These enveloping systems provide the reinforcement needed to make a feedback loop self-sustaining. These characteristics are the recognizable features of animated conscious life.

Furthermore, we can now see why we have been able to make good sense of the universe. Our thoughts are an integral, if small, part of the aethereal field; our thoughts (stimulated by experience) are more fundamental than the material world of experience that we observe. Our thoughts are as close as we can get to the underlying fundamental expressions of the apeiron.

Notes

[1] The material world is a deficient representation of the bigger aethereal picture.

[2] The book by Kaku and Thompson (1997) gives a good popular account of the search for a theory of everything.

[3] Occasionally a fifth element has been postulated to describe the spiritual and celestial realm or to describe an immaterial unity of the other four elements. However, the logic of a description of the indescribable is annoying.

[4] Arnauld is discussed in Stokes (2002) p 75.

[5] See Stokes (2002) p75.

[6] An introduction to intuitionistic logic is found in Iyanaga & Kawada (1977) 400G.

Chapter 13 : Philosophical Pictures

If the acknowledged role of science is to describe and to explain the quantitative phenomena of the universe, then the role of philosophy is to describe the noumena of the universe. Here, in defiance of some *philosophical* usage, I shall let the term *phenomena* refer to observable occurrences within an actual external physical universe (and not to those appearances which are devised and interpreted by our mental perceptions as such supposed occurrences) and in contradistinction I shall let the term noumena refer to processes stemming from mental-style activities (and not to some external processes occurring within a more fundamental[1] but essentially unknowable external world). The noumena may be generated as a response to some phenomena or some noumena. However, they may also be self-originating.

Consequently science's theories and science's principles are deemed to be noumenal. Now, a philosophical hypothesis is part of and attributes meaning to every scientific theory. However, not all of the philosophical hypotheses that describe the different contemporary scientific theories are mutually compatible. Nevertheless experience has shown that large sets of consistent philosophical hypotheses may be formed. A large set of consistent philosophical hypotheses (associated with accepted scientific theories) is known as a worldview or paradigm of science. Unlike scientific principles that must provide a simplified subsumption of a set of theories, a useful paradigm does not have to be more elegant than the sum of its constituent philosophies. Nevertheless we would expect greater elegance as we reach the ultimate scientific paradigm. Often there is a single mainstream scientific paradigm for an era. Yesterday we had two mainstream scientific paradigms (prevenient to eventual reality ie prior to and *heralding* the new theory), centred upon the theories of quantum mechanics and general relativity. In addition, there have always been heretical scientific theories that clash with the dominant paradigm(s). Within the context of these earlier paradigms, eventual reality should be classed as heretical.

Alternate paradigms can form around such heretical theories. A paradigm has been constructed around the hypothesis of eventual reality. In section 13.1, we shall discuss this paradigm and then show how the paradigm justifies all of existence (section 13.2). However, philosophy is not restricted to being an adjunct to science. Philosophy exists as a distinct discipline that engages in various forms of internal rational debate. Thus the hypothesis of eventual reality has implications for the discipline of philosophy and some of these implications are discussed in section 13.3.

CHAPTER 13

13.1 Fabricated Storeys of Reality

On planet Earth, the air we breathe and the water we drink have been recycled many times before we partake of them. It is equally difficult to find sensible ideas on the nature of reality that have not been digested previously by many other minds. Despite relative ignorance about the quantitative behaviour of the universe, our learned forebears recognized certain inescapable logical necessities for a self-consistent description of nature. Without the framework of modern science, our forebears' exploration of the early philosophical ideas may seem primitive to us but their core insights and their deductions were not naïve for their time. Consequently from the ancient Greek philosophers Thales, Parmenides, Aristotle and Plato to later European philosophers Spinoza, Leibnitz and Bishop Berkeley, the philosophical elements embedded within the theory of eventual reality have already been argued for thousands of years. Unfortunately the development of these philosophical ideas was marred by non-relevant historically contemporary philosophical concepts. Nevertheless we may use the legacy of their deliberations to uncover the self-consistency of the foundations of eventual reality.

Therefore let us now consider this multi-storey edifice of reality and probe how it can be justified.

13.1.1 The Primordial Constructs

Eventual reality provides a nested set of pictures of reality. The pictures in this hierarchy are embedded in the storeys that comprise the edifice of reality. We shall show that the taxonomy of these storeys partially resembles that classification (substrate, qualified, disposed, relatively disposed) given by the Stoic philosopher, Chrysippus, in the third century BC.

From a fundamental viewpoint (foundation storey, substrate), the universe is an infinite networked system built from a single elemental type: the event. The pointer qualities of an event identify the qualitative nature of the event; there is no event without these qualities. Thus events do not exist without the network of events and vice versa. The individual events (of this fabric of events) exist outside of any structure of space and time; indeed space and time are amongst the derived qualities of the next storeys in the hierarchical picture. Thus there are no temporal beginnings or endings (and no spatial cuts or limitations) in this picture of reality. This description of the grounding viewpoint delivers a picture of eternity without time and a picture of boundless immensity without space. This foundation network is a disconcerting balance between unbridled opportunity and unfathomable chaos.

The idea that there should be only one universal form, dates back at least to the seventh-century-BC philosopher, Thales of Miletus, who attributed all aspects of the natural world to the qualities of a single form, which he mistook to be that form exhibited by water in its many phases. Parmenides of Elea argued that the ultimately real system, displaying this unique universal form, should be spatially unbounded and indivisible. Heraclitus argued that the universe was in eternal conflict between incompatible warring opposites. Furthermore, a hierarchical structure to reality was accepted long ago; Anaximander argued that an infinite regression of storeys in this hierarchy was unacceptable, not every form could be founded upon some other form from a lower level. He surmised that there needed to be a foundation storey of reality and this foundation storey took the form of a borderless 'apeiron' (corresponding to our network of events) out of which everything else was built.

Such views of the ultimate reality became unfashionable but, in the seventeenth century, Benedict de Spinoza revisited the idea of a single primitive type. Spinoza argued that everything in the universe is one primal substance (nature or possibly God) that possessed many potential attributes. Adopting an apparently contrary viewpoint to Spinoza, Baron Gottfried Wilhelm von Leibnitz proposed that nature was comprised of an infinite number of discrete immaterial entities, which he called monads (a term first introduced by Plato). Physical objects are formed by the arrangement of these immaterial indestructible non-spatial non-interacting monads. Thus Leibnitz's monads share some (but not all) characteristic features of the pointer-events found in the theory of eventual reality.

Now, expanding upon these historical arguments of natural philosophy, it clearly makes sense that eventual reality possesses an ultimate foundation storey of reality. Just as we argued for a start to time in our hybrid-coordinate world, an infinite regression of derivative storeys would be paradoxical. In an infinite hierarchy, there are no special reference storeys from which to derive any unique features. How could the storey on which our consciousnesses reside acquire any distinctive characteristics, how could that storey be differentiated from those immediately above or below it? Thus the edifice of reality must have a primal storey of *first cause*.

The primal storey of reality is characterized by the absence of a lower justifying storey. Thus at the lowest level there can be no externally supplied inputs. Logically, at a primal-level description of reality, the count of available form types and the count of expressions of each form can be only a number which intrinsically possesses a special meaning ie one of the particular numbers 0, 1 and ∞. Any other non-special integer would require a stronger justification of its value from outside the storey. However, these three special numbers can have a rational justification[2] based upon what they directly represent.

Let us consider the possibilities. A zero count (of the number of expressions of a particular form) signifies a dismissible lack of contribution to reality. Alternatively, if there were an infinite number of different types, then there should be way to generate the distribution of types. Hence, we would be able to re-categorize the set of all types as a single monadic form for which the previous types were specialized forms. Thus it is justifiable to expect only one primal form (ie one elementary type) of something on the foundation storey of reality. Next, let us consider the number of expressed examples of this unique primal form. Clearly we dismiss a null universe provided by no examples of a unique primal form. Equally a universe containing a single example of a single form is also inconceivable. The primal forms are nothing without their qualities. However, these qualities cannot be imposed from outside the level. Hence, these qualities can only arise from the relationships between the primal entities. When there is only one primal entity, there will be no relationships, therefore no qualities, and no existence. However, we do not need to explain an absence of control as exhibited by an infinite repetition of examples of the unique primal form.

Thus we conclude that the foundation storey of any universe should be describable as an infinite number of examples of a single primal form ie an infinite number of monads. Each monad has a relationship with other monads; the ensemble of monads forms the boundless apeiron. Furthermore, each monad acquires all of its limited properties from its relationship with other monads. Now, in the absence of an external reference scale, the relationship between monads cannot be quantitative. Equally the relational qualities are *entirely* qualitative, the qualities cannot be based upon semi-quantitative relations such as bigger than or less than.

We may deploy a simple procedure to uncover the only natural form for such a qualitative relationship. Take a more conventional fictitious interactive relationship between monads, strip away all quantitative aspects of this relational interaction and that which remains is the essential qualitative nature of a relationship between monads. This distilled relationship may be identified with the pointer functioning that we have attributed to the event.

Hence, from simple *a priori* arguments we are able to arrive at eventual reality's foundation picture of reality – an infinite network of pointer events. However, the presented arguments have used the justifications and language of a higher storey of reality. It is worth remembering that a primal event is neither a location, nor a thing nor a process and that logic is not part of the ground storey of reality. These issues will be discussed again in the next section.

13.1.2 Coherent-Storey Picture

The primal-level picture of reality shows an infinite jumble of options. A discussion on how the options can be set is left for the explanations provided by the picture of a higher-storey. Next, we build the first upper storey that will display the coherence of the network of the primal level of reality. We are unconcerned with the incoherent aspects of the primal jumble because these incoherent aspects collectively nullify themselves. Identifying the coherence of the structural arrangement of networks serves both to define the allowable symmetry and to provide a means of establishing a quantitative scale for the coherent residue. The scale length is provided by an appropriate count of sets of events in the most symmetric (dominant) sub-network. Thus we have argued (section 9.2) that the view provided on the first coherent storey ('qualified' storey) of reality shows a multi-branched hierarchical structure. The base level of this structure is referred to as coordinate space.

Thus the base-level set of eight branches of this hierarchy defines our quantitative space-time. Actually each base-level branch displays a spatial geometry with eight coordinates that are related to the familiar space-time and momentum-energy variables. Further pointer links that do not contribute to the ground-level space-time represent structural stubble embedded in this space-time and this stubble gives rise to the aethereal field or aether. This aether is only required to exhibit local symmetries. Structural arrangements (for this aether) that display local symmetries will dominate over the fewer arrangements showing symmetries that are more far-reaching. Thus, in the picture belonging to the first coherent storey, the upper symmetry levels maintain their qualitative character whilst being subject to local symmetry requirements. Meanwhile, the stationary ground-level sub-networks provide a quantitative coordinate space.

The realized qualitative features of the upper-level aethereal structures are a form of language. Essentially the aethereal field is the collection of writings in a universal langue distributed throughout space-time. Much of these writings may be randomly generated and (like grunts and groans) such writings contain no important linguistic meaning. The symmetries and grammar of this langue deliver meaning through relationships and the foundation of logical principles. Thus the Platonic[3] ideals of mathematics and logic are found within a descriptive space and time without any matter.

By using classical logical arguments, a logic system is seen to arise naturally in the context of this first coherent storey. Clearly this argument is circular; logic is used to derive the need for logic. However, the character of the derived logic system is not a copy of the more restricted input logic. The circular (by substitution) argument shows that our logic-system is self-consistent. Any such self-consistency reflects part of the necessary symmetries of the universe that are expressed in the coherent arrangement of events. We also used logic to explain the characteristics of the grounding storey of reality. However, this explanation of the grounding storey was constructed from the viewpoint of our experiential storey of reality – this storey provides the logic that we understand. Thus the association between the pictures of the different storeys is part of the self-consistency that reflects the symmetries of the storeys.

Now, in this coherent-storey picture, the aether fully describes all aspects of nature. The precursors of matter are found as linguistic expressions within the aether and the dynamics and properties of matter are revealed in the next material storey of reality. Any non-random 'willed' expressions within the aethereal field are termed thought. The universal langue is available to be expressed throughout space and time; the universe is predisposed to expressions of thought.

13.1.3 Matter-Storey Picture

The coordinate space of the previous picture of reality is unbounded. This coordinate space retains some of the iterative features of the apeiron – it endlessly repeats the sequence of eight dimensions and eight branches into the aether.

Now, the edifice of reality is built through shedding non-analyticity of the aether and focussing upon collective and cooperative descriptions. Thus a new picture arises from an analysis of the collective behaviour of the aether. This cooperative/collective conduct is revealed by a reciprocal transformation that 1) converts coordinate space into hybrid space and 2) populates hybrid space with matter obeying quantum principles. When interpreted with appropriate units, this hybrid space is identifiable with space of our empirical awareness. Whilst the transformed aether of hybrid space is a linguistic thought field, observed matter is a contraction (across all the same time-sense branches of space) of the statistics of this linguistic thought field.

Hence, matter and thought are intricately connected. Thoughts and matter are each constructs of a lower storey of reality, yet there is a mechanism for the mind (a writer of thought) to receive stimuli from matter and to control the motion of matter. Thus eventual reality has a clear answer to the mind-body problem - a problem which has intrigued many philosophers. The traditional issues of the mind-body problem are simple: Which of the concepts of mind and body are more fundamental? How are mind and body to interact? Why do mind and matter obey different rules?

An eighteenth-century philosopher, George Berkeley, the Bishop of Cloyne was an early proponent of some relativistic ideas and, in addition, Berkeley promulgated some intriguing mind-body views that share similarities with the philosophical stance of eventual reality. Berkeley proposed that minds and ideas were more basic than brains and matter, which were to be borne solely upon ideas. Berkeley stated that we interfaced the material world through our minds; thus it was reasonable to assume that there are minds. The material world of our perceptions contains some of our mental thoughts but there is no proof that there is any further real substance behind these thoughts. Then he pointed out the futility of trying to prove that the *existence* of any matter that is independent of our own minds, let alone, any matter that is independent of the mind of God. Thus he concluded that the only reasonable assumption (despite appearing contrary to most folks' common sense) was that matter was not fundamentally real; matter had to be derived from thought since thought was all that Berkeley was prepared to declare as real.

More recently, the nineteenth century psychologist and philosopher, William James[4] took these ideas further and promulgated a doctrine of neutral monism in which mind and matter are both derivative of the ultimate 'pure experience' of fundamental reality; abstractly this 'pure experience' resembles the universal expression of the choices of pointer-events. These choices build the ultimate elemental experiences. Eventual reality talks of options and choices in the apeiron because those freedoms of choice and changes in time that we observe in the higher storeys of reality are correlated to an interpreted restructuring of the apeiron. Even options and choices are somewhat poor terms to use; our language is ill-equipped to describe the apeiron, although we have every reason to believe the apeiron is effable. However, the terms options and choices are preferable to abstruse phrases like 'self-stabilization of the composite structures around dominating symmetrical networks represented by fickle pointer assignments'.

13.1.4 Higher Storeys

There are many more storeys of reality; however, these higher storeys eventually become less distinct as the storeys get further away from the fundamental storey. The next two storeys owe their identities and characteristics to the certainties that arise from the repetition and aggregation of matter. The flimsiness of the quantum particles of the previous storey is tamed by the statistics of massive aggregations of matter, aggregations that can acquire some semblance of (classical) definiteness and certainty. A mixture of one or more spatial regions with quantum flimsiness and one or more spatial regions containing massive aggregations of matter identifies the next quantum-measurement storey of reality. The correlation of classical certainty with quantum uncertainty delivers the rules of quantum measurement. The creation and storage of memories require measurement processes. Hence strictly the operation of our minds spans this quantum-measurement storey and the previous matter storey.

The next storey (beyond quantum measurement) is the base storey for classical mechanics. In the base classical-mechanics storey, quantum flimsiness plays no part. Everything is dominated by massive localized aggregations of matter. These aggregated lumps of matter contain great numbers of particles that are in close proximity; the aggregations form across all eight branches of hybrid-space. Classical reality is deterministic, time-reversal symmetric and space-reversal symmetric.

The higher storeys become less distinctive and these storeys have often previously been considered as part of the classical picture of reality. This system of storeys is summarized in table 13.1.

Table 13.1

The multi-storey edifice of reality which is built upon the entirely qualitative apeiron.

#	Description	Ownership
5-	Cooperative assemblies	Plasma/thermodynamics, less distinctive storeys
4	Classical assemblies	Classical mechanics
3	Mixed assemblies	Quantum measurement
2	Hyperspace	Quantum principles, matter
1	Coordinate space	Mathematics, logic, space-time
G	Apeiron	Bountiful possibilities

13.2 Jus Naturale

Now, let us reconfirm that the hypothesis of eventual reality has a clear *a priori* rationalization. Recall that eventual reality has been presented as a theory derivable from four tenets without recourse to any other principles. Rationales for each of these four tenets are scattered through this report and these arguments will be summarized in this section. These rationales allow us to claim these tenets as the universal *jus naturale*.

The foundation storey of reality is sourced from these four tenets. Now, we image that each storey of the edifice of reality is to have its own concepts and precepts. We do not have to presume that there is more than one storey in this edifice; however, we shirk from allowing there to be an infinite regression of storeys without a lowest storey. If there were to be no foundation storey, there could be no distinguishable concepts or distinguishable precepts on any particular storey, and perhaps the edifice of reality would have to collapse until a finite foundation was established. We also refuse to accept that there could be no storeys in the edifice of reality. Clearly, the denial of the universe's right to exist (even if there were no logic) would require a storey upon which to place the denial!

Premise 1: *Every possible option is expressed.*

Simply stated, within the fundamental storey of reality, there are to be no imposed regulations ie there is no speed-of-light barrier, there are not even any spatial dimensions. Any imposed regulation would need justification from a lower storey of reality but there are no lower (more fundamental) storeys. Hence, there can be no regulations: every possible option is expressed on this storey.

Premise 2: *The foundation of reality is qualitative.*

Yesterday's scientists have tended to be blinded by the success of science into believing that the world is quantitative. However, nothing quantitative that exists can be regarded as elemental since everything quantitative has both a quantitative value and a quality to which the value is attached. Although most familiar qualities are compound mixtures of qualities, qualities can be factored into their essential essences. A reductionistic foundation of reality can only be describable using such essential essences (qualia). Thus if an elemental foundation exists it should be qualitative.

Proposition 1: *The principles of natural philosophy are hierarchical in nature and built using Fermat's principle upon an increasing semblance of order.*

There may be one or more principles of natural philosophy. A principle that 'there are no other principles' would suffice if there were only one principle. A hierarchy (called the edifice of reality) describes the way we choose to rationalize these principles so as to make the universe more understandable. However, we do not have to assume that there is more than one level to any hierarchy. The universe is the way that it is; however we might choose to describe it. Hierarchies are convenient, if we can construct them, because we are programmed to look for order and certainty in nature.

Now, premise 1 tells us that our foundation storey of reality lacks order and certainty – everything is allowed, everything happens. We construct hierarchies for our convenience and understanding of reality only – a hierarchy will not change the ultimate nature of reality.

Fermat's principle is a statement of dominating contributions derived from logic and mathematics. We employ Fermat's principle to reveal the pleasing coherence, order and certainty, that is hidden within the discord and hence to construct the first few upper storeys of the hierarchy (see table 13.1).

Proposition 2: *All of reality (including those parts that we recognize as describing mind and body) is ultimately built from qualitative elements that reference and relate as pointers to the other elements.*

We have argued that there is no justification for any fundamental reality other than an infinite repetition of instances of a single universal type. One explanation for this assertion relies upon there being no possibility of a justification for anything else. There can be no argument for why there should be any finite number of *different* types. Without such an argument there can be no restriction imposing a distinction upon types of monads.

Thus there can be only one universal type. Each instance of this universal type must possess qualities (premise 2). However, these qualities are not assigned from a lower storey of reality; these qualities can only be derived from the relation of an instance of type with the ensemble of all other instances. Since there is one type, there must be more than one instance of this type; otherwise, there are no qualitative relations, no qualities and no instances. In fact, by premise 1, we cannot impose any finite restriction upon the number of instances; thus there must be an infinite number of such instances.

The essential qualitative essence of any relation between instances of the fundamental type is derived from the referential pointer function. Hence, we identify the universal type with the fundamental pointer-event of eventual reality.

These arguments, supporting the four tenets, provide the reasons for the claim that the tenets of the hypothesis of eventual reality are the *jus naturale* of the universe. However, there are consequences to this claim that the tenets are apodictic tenets. Severely our hypothesis admits no inputs from other sources. A critic may argue that such fundamental tenets, if self-justifying, are tautologies that impose no regulation and hence any conclusions drawn from the tenets are similarly without informational value. It seems that such a theory should only be able to predict other apodictic truisms. However, there is a way to avoid labelling the universe as a platitude.

Our hypothesis holds three other implied constructs that are directly challengeable statements ie statements that are *not* tautologies. These statements are statements about the theory; they maintain that eventual reality

1. is founded upon tenets that can be logically self-authenticated
2. has no possibility of further input (eg fudge factors or externally applied principles)
3. applies to everything without exception.

These additional statements possess the highest level of information content. Eventual reality is indeed a richly predictive theory. The science of the universe need not be insipid, though perhaps the universe remains tautological! However, we are simply exercising our prerogative to account for the universe in a rational coherent manner.

13.3 The Philosophical Discipline

There is no mainstream paradigm of philosophy; the discipline of philosophy has had no umpire to choose between its divergent schools of contrary thought. For instance, the unresolved conflict between two of the major schools of thought (the rational school and the empirical school) is well-illustrated by Immanuel Kant's antinomies[5]. Kant systematized these antinomies as unavoidable contradictory pairs of theses and antitheses. For example, Kant's first antinomy deals with whether there was a beginning to time in the universe. The rational thesis argues for a beginning and an evolving universe as discussed earlier in section 9.3.1. The empirical antithesis queries the sensibility of a starting time and asks how the universe could possibly spring into existence from a timeless state when timelessness implies immutability.

Now, as a theory of reality, eventual reality is able to act as an umpire for all valid philosophical arguments. Indeed, the clashes of perspective as demonstrated by Kant are to be expected. The arguments of Kant's sets of antinomies need to be identified with appropriate storeys of the edifice of reality. Eventual reality weaves a synthesis of the rational, the empirical and the righteously sceptical (coined here to describe insightful but *apparently* untenable beliefs). Generally inductive arguments starting from the empirical reality of hybrid space will lead to conclusions applying to a lower (ie more fundamental) storey of reality. Rational deductive arguments, which start from the precepts of the intermediate reality of coordinate space, will lead up to a higher storey. Finally only far-sighted and bold[6] eccentrics entertain the righteously sceptical declarations that describe the reality of the apeiron ie the reality of the foundation storey.

For instance, eventual reality recognizes that time has a beginning in hybrid space (atomic units); time is unbounded in coordinate space; there is no time in the apeiron. This adoption of every possible answer is not counterintuitive; reality is built from storeys each with its own framework of validity. These storeys support each other in the grand edifice of reality.

This multi-picture explanation is not new. Early Greek philosophers recognized that reality contained a hierarchy of contradictory pictures. However, the philosophers lacked sufficient knowledge of science. They were unable to establish concrete relations between the conflicting pictures. Historically their explanations have been criticized as mysticism. For instance, Plotinus[7] formalized a trilogy of pictures of the universe during the third century AD. Plotinus' first picture of the universe describes *the one*. *The one* is everything but it is timeless, spaceless and possessive of indescribable beauty. *The one* corresponds to Anaximander's apeiron or our network of events and, in our language, indescribable beauty becomes the duality of unlimited opportunity and indescribable confusion. Plotinus' second picture (first-elevated storey) of the universe holds *the intellect* of *the one*. This *intellect* is the light by which *the one* sees itself. In the language of eventual reality, the intellect corresponds to the coherence of the apeiron by which the apeiron self-adapts into an orderly realm of coordinate space with accompanying aethereal field. Thus Plotinus' light is our coherence. Plotinus' third picture of the universe describes *the soul*. *The soul* is formed out of the intellect. Every aspect of the universe of our experiences is built from *the soul*. *The soul* is our aether. *The soul* has two parts: the inner part of *the soul* is responsible for the mental realm of experience. The outer part of *the soul* is responsible for the material universe. The *inner soul* corresponds to the linguistic expression of the aether on hybrid space branches; the *outer soul* corresponds to the statistics of the branch-contraction of the aethereal field in hybrid space.

Furthermore, eventual reality impinges upon the acceptability of much philosophical argument through the appropriate use of logic. Eventual reality requires four logical values for any statement; two of these logical values are decidable true and decidable false and two of the logical values are undecidable true and undecidable false. Hence we observe that the paradoxes of infinite recursions and infinite sets are related. For instance, we are to adopt an intuitionist view of Russell's paradox. One may prove membership of an infinite set but never prove non-membership of an infinite set since however many elements have already been compared with a queried object there will always be even more elements that have not yet been compared.

Next, we shall look afresh at how eventual reality answers some of the great questions of philosophy.

13.4 The Great Questions

Many of the intriguing puzzles involved in the great philosophical questions of fundamental experience are indicators of an unquestioning acceptance of an inadequate paradigm. Such difficult-to-comprehend questions seem to make sense and seem to demand an illogical justification or explanation. We shall dissect some of these questions in this section.

13.4.1 Why is the universe here? Could there be nothing at all?

From the perspective of eventual reality, let us explore why the universe is required to exist. We shall start by imagining an empty spatial universe. For a material monist, a completely empty universe has no reference scale and, hence an empty universe is unacceptable. Unfortunately, taken to their extreme, the ideals of material monism lead to space necessarily being regarded as a plenum. Intriguingly the eventuality of an empty (of matter) space is quite conceivable and achievable within the theory of eventual reality. An empty universe would require that every event conspired to form the highly improbable framework of a matter-free space ie every pointer link belongs to a common network making an $SU(2)/Q_8$ space. Clearly a hypothetical empty universe is an exceedingly special case. Now, we would expect there should be a strong reason for the occurrence of such an exceedingly special case. Alternatively we should not be surprised to observe matter in space since a space containing matter is what is to be expected of practically every conceivable universe.

A similar logical reasoning could be applied to the explanation for why there is a universe. A non-existent universe (pictured as every event choosing not to point) would be an incredibly unlikely special case. The general, most likely case is that case in which an infinity of events point to other events and hence these events acquire environmentally determined relational qualities (in addition to their pointing function).

Truly our universe is unique but the universe is without any special requirement; such a universe requires no justification simply because the inconceivable does not eventuate. However, this argument is awkward because we have a real difficulty with the concept of a non-existent universe that is constructed from non-existent non-pointing events. Clearly a universe can be defined, if a universe exists. Likewise, a non-existent universe can be defined only as what it is not when the concept of a universe (and the universe itself) actually do exist. Essentially the concept of nothing at all has no *a priori* relevance. Indeed a rule of non-existence of everything is a self-contradiction since at least a universe of one or more rules must exist to house this particular refutation. Such denials of lone negative concepts are part of the Parmenidian tradition.

13.4.2 What came before the start of time? How did the universe spring into being?

The foundation space of eventual reality is unbounded. Time, 3-space, specific energy and specific 3-momentum can take values from negative infinity to positive infinity in coordinate space. However, the atomic time of our experience is a mathematical-real root of the square of the coordinate time of this foundation space. Thus time as measured by our timepieces, starts with the big bang at time zero. There is no (atomic) measurable time prior to time zero. However, questions of ultimate existence belong on the foundation (apeironic) storey of reality not in the matter-filled world of our experience.

To explain the origin of time, let us imagine that we could move in coordinate time from positive time through the coordinate origin to negative time. Another observer recording atomic time would observe a temporal bounce, atomic time decreasing and then atomic time increasing again. This scenario is not unlike an explorer heading in a fixed direction (initially due south) on the surface of the Earth. The explorer reaches the South Pole and continues with the same heading that we observe is now due north. In both examples, there is a coordinate system in which the excursion is un-deflected and a measurement convention in which the excursion undoes a direction reversal.

In the case of the Earth, both pictures are valid because there is an extra dimension in which to view the travels. In the case of the universe, both pictures are valid because we have hypothetical alternative systems of units in which to represent the description of reality.

13.4.3 What is consciousness? What is free will?

Within the apeironic foundation storey, the universe possesses rather too-many options. However, the interpretation of every option is in relentless competition with the interpretations of all other options; the combination of almost all of the sets of arrangements of options effectively amounts to an irrelevance when viewed more macroscopically. However, certain symmetric sub-ensembles of pointer options seem to cooperatively conspire to create a combined coherence that represents a more macroscopic field of expressible choices imprinted upon a multi-directional multi-branched spatial framework that we call coordinate space. This field corresponds to a universal langue. The scope of these choices dictates the vocabulary of the langue of the universe. This immaterial linguistic field has several names. I have referred to the field as the aether or the aethereal field; the aether is the paint on the canvas of the mental realm. When viewed in experiential hybrid space, the values of the aethereal field become settable in atomic time (time expressed in atomic units) by using quantum processes including measurement. These fixed linguistic values in hybrid space are termed actualized thought.

Thus thought is a natural and basic constituent of the universe. Our consciousness is recorded in the linguistic field of the universe and not in the matter of our bodies. Of course matter is also a function of the aether. Thus our thoughts are as one with the aether. Free will is simply the freedom of the universe to express a thought. The format is constrained by symmetries but the freedom within these bounds is founded in the total freedom described by premise 1.

Now, our conscious awareness is also related to our ability to store and retrieve memories. Without this memory ability, we have no individual self and the freedom of expressed thought will have no continuity in time and no identifiable purpose. Hence most of the aethereal field is without expressed will.

13.4.4 How can we explain the passage of time? Why do we only remember the past and not the future?

The metric of general relativity describes a fixed space- time landscape (a 'timescape') in which past and future histories are displayed without any sense of temporal flow. Hence, many physicists have expressed doubt that empirical time flow could be real. However, the foundation coordinate space (of eventual reality) has time and position dimensions but also extra dimensions that describe a rate of change in time and change in position. Thus a measure of passage is inherent in the structure of the spaces of eventual reality. This measure of passage records the rate at which the (mathematically-imaginary) coordinate dimensions change. This relation between mathematically real and imaginary coordinate dimensions is dictated by the requirement of separability of the component symmetries. This same separability of symmetries is required for the aether. In the aether, the primal processes dictate the analogous change-rate of notions. Change rates are an essential feature of the formal structural of the universal langue.

Thus an acknowledgment of the flow of time is an inherent facet of eventual reality's spaces and of our thoughts and consequently of psychologically constructed time.

Local fluctuating choices can be set in time by a quantum measurement. Thus time is indelibly linked to the permanent setting of choices. The present is special; it is where the choices of our awareness are becoming set. Past choices continue to influence the future. Thus our thoughts ride the crest of the advancing wave of newly available choices. Our appreciation of this flow and our sense of the moment both come from our conscious role in setting these choices and our access to past choices.

Quantum mechanically the future is undecided and cannot be remembered unlike an absolute past that also potentially has impact upon the present.

13.4.5 Do we have souls? Is there life after death?

Eventual reality does not accept the primacy of matter. Matter is treated as a localized representation of a statistical system that acquires some observable (certain) knowledge. Certainly the theory holds that there is an essential immaterial part of every individual; this part is not fully confined to the spatial extent of the corporal body. This immaterial part hosts the arena for our thoughts including: emotions, beliefs, ideas and decisions. Specifically this immaterial attachment resides in the aether on the branches of spaces. The choices that we make thunder through the aether of surrounding space, we are enveloped by a disturbance wake in the aethereal field; this disturbance wake is triggered by our mental activities. An individual soul may be identified with this immaterial disturbance to the aethereal components that shadows us. The continuity of our mental awareness is contingent on our accessibility to our individual memory. Memory is held in matter and thus our essential existence (soul) is tethered to the corporal body and dependent upon energy degradation.

The death of our bodies ends our maintenance of these effects. The traces of our soul disperse and merge with the evolving universal collage of thought from all life of all previous time. Individuality and continuity are no more; our individual essences are no more.

13.4.6 How do we describe existence in conceptual terms? What is the essential nature of the universe?

We can often use the term existence without qualification. However, there are several different contextual meanings of the word existence. We shall consider an example. Physically animal horns are observed to exist (as material entities in our hybrid space). Likewise other descriptors: white, mammalian, living and horse-like are exhibited by physical things within the observable (hybrid space) universe. Thus a unicorn exists as a valid concept but a unicorn (as the unified physical embodiment of all the necessary characteristics) has never existed physically even though all of its constituent qualities are regularly observed in nature. This example illustrates that there are three important styles of existence: - a physical/material existence, a conceptual/aethereal existence and an eventual/apeironic existence.

These styles of existence are subject to different maxims. The separate physical existence of all constituent elements is not sufficient to imply that a general composite entity exists physically (eg a unicorn). However, conceptual existence of all constituent elements is sufficient to imply that a general composite entity exists conceptually (eg a unicorn). The eventual existence of everything in the apeiron is implied by the eventual existence of any member of any event. Furthermore, consider a special package of elements comprising the set of everything attributed to one hierarchical storey of reality. By definition, the existence of this set of everything on a storey is demonstrated by the existence of all of the elements of the set. This latter maxim overrides any contrary conclusion of a previous maxim.

Thus if matter physically exists, then a material universe *physically* exists. Whereas, Descartes' 'I think therefore I am' is a claim of *conceptual* existence and not a claim of *physical* existence. We may generalize Descartes' claim; for all styles of existence. This general form of the claim states that by showing that an indivisible (non-composite) entity has a necessary quality we have implicitly demonstrated that the entity must exist. The style of such existence is to be determined by the storey of reality.

Existence is a property or characteristic that pervades the universe. The apeiron exists if and only if the individual events that comprise the universe exist. The events exist if they have a quality. However, the qualities of an event are derived from the relation of the event to the entire universe. Since existence is a quality, we are faced with a circularity that commonly underscores many of the arguments of eventual reality. It could be argued that this circularity betrays the symmetry of a menagerie of all the logically conceivable and logically inconceivable universes. Logically inconceivable universes are indistinguishable and can never be proven to exist. Only the logically conceivable universe exhibits the property of existence. A fuller treatment of this subject was given in question 1.

13.4.7 Why should we accept that there is a reality beyond our mental perceptions? How much subjectivity do we unwittingly impose upon the view of reality?

Most scientists would invoke Ockham's razor and reject consideration of such questions claiming the questions are irrelevant and futile. It does not matter whether science describes a dream world fabricated from the mental perceptions within our heads (may be just one oversized solipsistic mind) or science describes a true external world. The world that science describes is the only world of our perceivable interest. However, these questions of objectivity have beset empiricist philosophers who hanker after an assertible truth. Thus I will attempt to elucidate further reasons why an empirically inclined philosopher should believe in eventual reality.

Exclusive schools of philosophical thought are almost inevitable when the philosophical schools generate radically incompatible viewpoints. With philosophical conflict so widespread, empiricists have been justifiably reluctant to accept any rational conclusions that do not have empirical supportive evidence. Thus some philosophers have come to accept that only science (and its successors) can arbitrate the confounding mysteries of philosophy. Now, the hypothesis of eventual reality has evolved from accepted theories of science. Indeed the foundation of the hypothesis (of eventual reality) is based upon scientific empiricism (see part 1) and follows scientific traditions. The hypothesis acknowledges scientific observation and testing as the arbitrator of its validity. The hypothesis makes scientifically testable predictions. Hence, eventual reality holds the prospect of the theory becoming the accepted progenitor of all scientific theories. Furthermore, eventual reality eclipses the existing schools of philosophy. Eventual reality (and its successor theories) can accommodate all genuine disagreements by recognizing that the differing conclusions relate to different storeys of reality. All philosophers may look to eventual reality to heal their philosophical disagreements. With the circumvention of these philosophical disagreements, eventual reality can be held to be true (ie be shown by science to be true). Eventual reality reveals an existence of a reality beyond our perceptions. If eventual reality is correct there is an external reality.

Now, it is likely that every philosophical schema[8] has been unwittingly tainted by many aspects of the culture of the era in which the scheme was developed. Yet it seems that each new schema presents itself as exempt from any cultural influence. Eventual reality does not deny this criticism of subjectivity ie being unintentionally influenced by both the previous scientific paradigm and by societal culture. Eventual reality admits no brakes in its coverage of reality; every attribution and assignment involves interpretations of arguments and equations. There are countless attributions and identifications within the hypothesis as presented here. Undoubtedly many of these attributions have been tainted by subjectivity and the attributions will need to be refined and clarified. In particular subjectivity is pertinent to the many circular arguments that confirm the presumed status of a solution of an analysis without ruling out more useful alternatives (if they exist).

13.4.8 Why do the parameters of the universe appear to be finely adjusted for life? Could there be universes with different laws?

Clearly much of terrestrial biology requires: sunlight; reasonable abundances of carbon, hydrogen and oxygen; non-oppressive gravitational fields; planetary ionospheric radiation shields; punctuated environmental stability over evolutionary time-spans; etc. However, these are not the actual requirements for life. Even the variety of terrestrial biology shows that most if not all these requirements are unnecessary.

The seeds of intelligence are programmed into the aether. For small planet-bound sapient life instances (such as humans), memory needs fixing by quantum measurement and preserving by expenditure of energy. However, the model of a human does not define the requirements for life. Intelligence may be rife in forms of life, yet life does not need intelligence nor, in principle, does life need common attributes such as reproduction or responsive motion. Thus let me give a different definition of life. Life is something that is alive and spatially bounded by a single surface. Alive means being an active member of a self-regulating ecosystem and directly participating in the reprocessing of matter (eg oxygen, water) as part of the ecosystem's cycles.

The universe is finely tuned to the requirements of biological life-as-we-know-it simply because we are finely tuned adaptations to the universe. Anthropic evidence reflects our adaptation to the universe, not the universe's adaptation to or selection based upon, our needs.

The physical laws and parameters are a result of a lack of applied will. Fixed parameters cannot be changed since they are the expression of the massive free indecision predominating in the universe. Hence, there are eight dimensions on each branch of coordinate and hybrid space. There cannot be nine or seven dimensions. There is no capacity to manufacture a universe that meets our requirements. However, there is a capacity for life to adapt to the universe and even for life to select a storey upon which it can evolve.

13.4.9 How can there be change if the apeiron is timeless? How can there be individual things if the apeiron is spaceless?

Change is a feature of processes that occurs within time, individualization is achieved within spatial dimensions. The ancient Greek philosophers accepted the argument that there couldn't be any change in the timeless origin-of-all-things and likewise there couldn't be positional differentiation in the spaceless origin-of-all-things. These conceptual arguments were taken up by Christian religion and incorporated in the notion that the Almighty Creator was immutable and impassible.

However, we must investigate what is meant by the key words *timeless* and *spaceless*. In common usage, there are several possibilities that we must distinguish.

Suppose that timeless is to mean *eternal*, then the apeiron is to exist within time, for all of time. There can be no constant decay or constant growth; there is no reference from which to allow regular behavioural changes and there cannot be irregular changes (these might lead to an accidental loss of eternal existence). Thus there can be no change in anything eternal. Mathematical truths are eternal. The discipline of mathematics grows and thus the discipline is not eternal.

Suppose that timeless is to mean *exterior to time* ('extemporal'). Then time also exists but we observe time from a vantage point beyond any time value. Figuratively we look down on the timescape (eg space-time continuum of general relativity) which shows all of time lain out before us. There are changes in time but there are no changes to the timescape (unless the observer descends into time to instigate a change). Thus we can form an extemporal view of the timescape of a physical theory like general relativity.

In eventual reality, timeless is to mean *without time* ('atemporal'). This concept is made clear by the picture of the apeironic network of events. Our network of events has endless options that logically exist without any reference to time. Imagined conspiracies within these options offer the promise (in a higher storey of reality) of a derivative time, of changes in time and of free will. However, none of these derivative concepts feature within the ground storey of reality! On the ground storey, there is neither change nor constancy. The timeless character transcends time and temporal concepts such as change and constancy. However, on higher storeys of reality we invoke the notion of time and change.

Equivalent arguments exist for different potential attributions of meaning applicable to the word spaceless (ie position-less). Spaceless could mean without any positional boundaries, it could mean exterior to or outside of space or it could mean without space. In eventual reality spaceless means without space; the apeiron does not exist within a spatial structure. The positional dimensions only exist on the elevated storeys of reality.

Thus the picture provided by the network of events shows that the timeless and spaceless attributes of the apeiron cause no logical problems or restrictions. These attribute genuinely refer to systems without any conceptual time or position.

13.4.10 How can we claim knowledge and understanding of an underlying reality not amenable to experimental confirmation? How can an ultimate self-justifying storey of reality possess any non-trivial content?

There are no isolated storeys in the edifice of reality. Each storey of reality is interconnected to one[9] or two other storeys. The more microscopic processes on one storey affect the occurrences on the next higher storey. We have seen that the foundation storey can be justified on rational grounds. Indirectly, because the storeys are connected, the non-contingent features of our observable universe also confirm the acceptability of the hypothesized apeironic form from which they can be deduced. Then we appeal to Ockham's principle. With the rational justification of the lowest storey and empirical justification from higher storeys, it is only reasonable to accept all the conceptual details of each storey as being justified. For the apeiron, any bias (springing from our limited natural senses), which creeps into the picture, will only assist us in the picture's application and understanding.

Now, the apeiron is indeed both trivial and momentous; such opposites are inseparable. Similarly order and chaos are one virtue of the one whole. Strangely opposites are not self-contradictory. The apeiron is self-authenticating without impositions that could restrict the nature of universe. Yet, trivial modes of description are without merit to us; they are content-free. Thus we seek those pictures of reality that are conceptually rich in derivative content. We select for relevance (non-triviality) and for regularity by employing Fermat's principle. Consequently we are responsible for our dismissal of the unwanted trivial descriptions and the selective acceptance of an interesting orderly description of nature. Nevertheless, in some sense, these orderly descriptions (eg the derivation of quantum mechanics) are practical tautologies ie the only logically acceptable (non-trivial) laws of nature. In practice the observed exhibition of these laws by the universe displays the behaviour of contingent information. The reader's lifespan and the mass of the Earth represent important examples of the rich spread of contingent information content that is unrestrained by a content-free foundation. Thus as many of us have suspected and tried to teach our students, all the laws of science are obvious.

Notes

[1] Noumena are the phenomena of the aethereal world. The aether (of hybrid space) is not more fundamental than matter since both occur on the same storey of reality. However, the aether is richer in capacity; matter is simply the statistical froth floating on the aethereal brew.

[2] Only these three particular numbers satisfy indiscriminate algebraic rules eg $x+0=x$, $x\times 0=0$, $x\times 1=x$, $x+\infty=\infty$ and $x\times\infty=\infty$ applying for any x other than these three particular numbers.

[3] In the tangible world, no line is ever perfectly straight; no sphere is ever perfectly round. What we encounter are approximations to beautiful idealized mathematical forms. The universality of these forms and a hankering after examples of spiritual permanence, lead Plato to postulate that these mathematical forms exist in another kind of world.

[4] See discussion in Russell (2004) pp 723-729.

[5] These antimonies are discussed in Sorensen (2005) pp287-302.

[6] There is a certain boldness in presenting seemingly perverse views. Many crazy and deluded people promote theories of everything, miracle cures for all ailments and eternal-motion machines. Now, when I started this work I did not expect that I would uncover an ultimate reality; if I had believed this then I should have been judged crazy (until proven sane). However, I did believe that I was as capable as anyone else of uncovering the next major step in the development of science. Hence I wonder whether I could be deluded (until proven rational).

[7] See discussion in Russell (2004) pp 269-280.

[8] This issue has been discussed by Russell (2004) pp709-710.

[9] The foundation storey of reality is the only storey connected to just one storey. However, we might easily roll all the classical storeys into one storey and make this storey the top of the edifice. Then we would have two singly connected storeys.

Chapter 14 : The Case for Eventual Reality

Eventual reality has a unique paradigm in which microscopic confusion builds hierarchical macroscopic semblances. Each storey becomes more regular, more definite and yet it becomes less fundamentally real than the previous. Familiar physical laws such as conservation of energy are regarded as macroscopic illusions. Experiential space is a transformation of the more fundamental coordinate space. Coordinate space is formed from combinations of only two group symmetries. These symmetries unambiguously fix the dimensionality and branch structure of space. Coordinate space is filled with the aether, the stuff of everything that we know of currently – both mind and matter. The aether is qualitative but ascribed quantitative statistics and it possesses the same pair of symmetries. A localized macroscopic view of the order within the aether yields an explanation of matter and of quantum mechanics. Yet the aether expresses both process and form. The aether can accommodate matter and the structure of conscious mental processes.

In the search for an understanding of reality, this paradigm is a major achievement of the theory of eventual reality. Further it has been argued that this paradigm is sourced from self-evident tenets. The value of the paradigm is evident in the provision of a resolution for countless philosophical wrangles that have arisen throughout documented history.

The value of the proposition called eventual reality will be considered in several sections. Firstly we will look at the limitations of the current paradigms of science. Then we will summarize some of the enhanced scientific understandings that the theory of eventual reality provides. Finally we will list the few genuine deductions that the theory has already made.

14.1 Elegant Paradigm

Elegance is one measure of a paradigm. An accepted aim of science is the move towards an ever-decreasing number of fundamental principles. Eventual reality has only self-evident principles.

The old worldview contains a disappointingly large number of principles. Indeed there are two accepted twentieth century worldviews: one for the large scale (based upon general relativity) and one for the small scale (based upon quantum mechanics).

Firstly we will consider the quantum-mechanical worldview. Let us ignore the fact that quantum mechanics is internally inconsistent and that it is poorly compatible with relativity. A mechanics only deals with the motions of materials things. Thus quantum mechanics, or any other mechanics, does not explain or describe space. Space is inserted in the theory as a postulate. Similarly in a mechanics, it is presumed that matter in some localizable form (whether that of a particle or of a higher-dimensional object) exists. Thus the existence of material constituents is another given concept unjustified within the quantum theory. Furthermore, quantum mechanics is subject to other long-held principles, these principles include causality and a form of determinism.

Next, let us consider general relativity. General relativity has a purer worldview than quantum mechanics. However, general relativity suffers from the similar problems and accommodates arbitrary assumptions (eg the form of its algebra). General relativity deals with the dynamics of space and gravitational fields. Indeed general relativity's handling of the discrete matter/continuum-space duality is the theory's greatest achievement. However, the theory is unable to account for the intrinsic properties of matter. Furthermore, this union does not dictate the dimensionality of space or provide a scale for the masses of elementary particles.

In addition general relativity remains steeped in the philosophical dogma of classical mechanics: causality, determinism, realism and absolute precision. These principles are contrary to free will and contrary to quantum uncertainty. Thus general reality is severely confined to the large-scale (cosmological) scenarios or to applications involving strong-gravitational fields. Further, general relativity cannot select between rival large-scale cosmological models. These cosmological models add baggage to the large-scale paradigm of general relativity, which is no more simple or unencumbered than the small-scale paradigm of quantum mechanics.

In contrast, the claimed scope of eventual reality is unrestricted – although this scope has not been completely demonstrated. Supposedly eventual reality can explain everything: space, matter, the quantitative, the qualitative, a mental universe and most importantly it can explain itself. This complete comprehensiveness is why eventual reality has been labelled 'a reality'. This complete comprehensiveness is why the theory of eventual reality is a theory of natural philosophy, not a theory of any one branch of science.

14.2 Explanations of Physical Quantities

The vision of eventual reality has allowed the new hypothesis to answer questions and to provide explanations for many physical principles and properties that previously were without an explanation of their origins. Many explanations have been presented throughout this part of the work. Some of these explanations will be briefly summarized here to illustrate the scope of current explanations already exposed.

Chapter 14

14.2.1 Existence of Scientific Laws

All of nature's regular rules (scientific laws) reflect precise symmetries[1] that are found in the coherent remains of overabundant opportunities inhered in the microscopic disorder. The correlations that emerge as a coherent consequence of the large-scale competition of every possibility become the accepted basis for a representation of reality displaying quantitative rules. As each storey is built using Fermat's principle, we see more and more definite order in each macroscopic description. The laws of nature epitomize this reified order.

14.2.2 Dimensionality of Space-Time

The dimensionality of space is determined without ambiguity. Only two symmetries are allowed. These symmetries are those symmetries displayed by mathematical-complex numbers and quaternion numbers. Space is made from the intersection of subspaces with these symmetries. Thus the reduced representation of the foundation space (known as coordinate space) has four explicit dimensions displaying a vector arithmetic based upon quaternions. However, each of these dimensions is mathematical-complex combining position and conjugate velocity as a mathematical-complex vector and combining time and specific energy as a mathematical-complex scalar number. Furthermore, these intersecting subspaces belong to a set of eight branches (displaying the same mathematical-complex and quaternion symmetries) that are occupied by an aethereal field, which again displays these ubiquitous symmetries.

14.2.3 Continuity and Uniformity Principles

The apeiron does not exist in a positional space; indeed the apeiron has no dimensions. Thus the apeiron can have no sense of spatial continuity and no sense of uniformity[2]. Likewise, on its smallest Planck scale, the foundation stationary space (coordinate space) and the stubble that we form coherently out of the apeiron is neither smooth nor uniform. Nevertheless, on much larger scales, the mathematical process of averaging (over large numbers of interconnected events) ensures that all aspects of physical reality will appear to be both continuous and smooth in both time and positional space. Examining the universe on different macro-scales provides us with rich interpretive descriptions of reality. Smoothness in quaternion-vector dimensions creates the impression that the quantitative descriptions of physical reality are spatially continuous and mathematically differentiable. Hence our macroscopically derived laws of nature are analytic. Likewise smoothness in the time dimension ensures that physical experiments are not influenced by when the experiments are completed; thus those laws, which were consistently obeyed yesterday, will continue to be obeyed tomorrow. Note that this latter conclusion (uniformity in time) is needed to explain the functional usefulness of the scientific method of induction which is very important to the scientific method.

14.2.4 Determinism and Causality

Indeterminism and acausality are not processes that require a justification; however, the mechanism that turns this fundamental holistic freedom into local determinism and into local causality does require an explanation. On a particular storey of reality, the reductionist stance is to describe quantitative reality in terms of a multitude of constituents that exist discretely in space and time. Until we climb to the classical storey, individually these constituents will not consistently obey determinism or causality. Nevertheless the unpredictability of individual constituents is washed out when we examine the behaviour of sufficiently large clusters of these constituents. Statistics requires that the macro-clusters behave in a predictable manner leading to apparent determinism.

Aspects of this predictability of the behaviour of the cluster may be transferred to another constituent of reality with which it temporarily interacts. In this way we are able to causally engineer our futures. When we reach the classical storey, all macroscopic features are overwhelmed by the statistics of constituent elements. These macroscopic features appear definite and determined by their interactions with other similarly definite features.

14.2.5 Phase Space of Statistical Mechanics

The phase space of statistical mechanics defines the dimensional freedom within which, the available to matter can be distributed. Now, hybrid space is eight dimensional with six quaternion-vector dimensions and two quaternion-scalar dimensions. The collective transformation of the aethereal field reveals matter and this transformation only involves the quaternion-vector dimensions. Furthermore the imposition of classical ideology has hidden the explicit mathematical-complex symmetries from the formulation of statistical physics. Thus these mathematical-complex symmetries are only completely revealed within the theory of eventual reality. The degrees of freedom exhibited by phase space and the equivalent treatment of the mathematical-real and mathematical-imaginary dimensions are a natural consequence of eventual reality.

14.2.6 Second Law of Thermodynamics

The concepts of matter and entropy are intricately entwined. Matter appears on the second-elevated storey of reality because Fermat's principle requires that we seek a stationary decomposition of the entropy parameter. Thus the second law of thermodynamics[3] is an offshoot of an application of Fermat's principle. However, like all laws of nature, the second law of thermodynamics does not apply on all storeys of reality. The second law applies to space and matter; this law does not apply to the opportunistic chaos of the apeiron.

14.2.7 Quantization and Discreteness

Little information about reality is available unless we resolve the picture of the universe into regions of space and time. The more localized we make such differentiations the richer the description of reality. On the storeys possessing matter this procedure generates the mathematics of particle-state creation and annihilation operators. Hence the mathematics introduces the concepts of quantization and discreteness in response to the need for a spatially discriminatory collective description of physical reality.

14.2.8 Compatibility of Quantum Measurement Rule and Time-Evolution Principle

Scientific reality is a construct built from many storeys. Each storey has unique concepts and unique precepts. Thus a principle like determinism can hold on one storey but not another lower storey. The incompatible principles belonging to the different parts of quantum mechanics are due to these parts being on different storeys of reality. The storeys connect the parts of quantum mechanics; the parts of quantum theory only seemed incompatible because of the way they were described.

Now, the principles of quantum mechanics are described by mathematical-complex algebra. Measurement is described by mathematical-real algebra. This reduction of the mathematics arises from the symmetries imposed by the measurement boundary conditions.

Likewise the quantum principles appear to be deterministic. However, interactions mix the states propagating on different the branches. In particular, the movement across branches with different time sense renders the quantum processes ultimately indecisive ie non-deterministic, yet quantum processes remain describable in a time-sensitive relativistically causal way.

14.2.9 Quantum Measurement

Any spatial or temporal disturbance causes mixing of the aethereal field on the different branches. Measurement instruments possess classical variables that interact with quantum systems to create strong inter-branch crossings. The measurement process is a temporally recursive procedure trapping a system between strong inter-branch connections. The recursion loop produces indeterminacy and potentially out of light-cone connectivity. Hence the recursion loop adapts causally as the measurement process goes through to completion.

From eventual reality's interpretation of quantum mechanics, the traditional measurement paradox is not paradoxical. The greater isolated system consisting of a measurement apparatus, a study system and an equipment set-up phase evolves indeterministically in time thanks to the inter-branch mingling. Indeed there is never any absolutely precise determinism even in the classical realm.

14.2.10 Symmetries of Fundamental Particles

Eventual reality explains the existence of fundamental particles, the formation of bosons and fermions, the number of generations of elementary particles, the electrical and colour charges of elementary particles, etc.

Particles are a product of a local picture of reality. Particle symmetries and quantum numbers are derived naturally from the two fundamental division-ring symmetries. For instance, the labels of the aethereal branches are hadron colours; the count of quaternion-vector branches is proportional to electrical charge.

14.2.11 Algebrae of Quantum Matter

Eventual reality is an *ab initio* theory of reality. It builds the storeys of reality from founding principles. The first elevated storey of reality describes coordinate space. Now, the quantitative reality of coordinate space is expressible in factorable mathematical-complex quaternion numbers. However, the process of extracting a local-style reality changes the dominant number scheme. This extraction process initially introduces an extension to this algebra and subsequently the algebra of the aether is transformed into a Grassmann algebra that delivers a particle-based description of reality with suitable families of both fermions and bosons. The principles of quantum mechanics apply on this matter storey of reality. Whereas the derivation of quantum-mechanical principles probably supports any model based upon a tumultuous underlying reality, the additional rationally argued number forms required in the derivation are unique to eventual reality. These number forms strongly support the underlying symmetry of coordinate space.

14.2.12 Actuality of Wavefunction

The wavefunction is a genuine mathematical-complex probability distribution function for quantum-particle fields. This distribution function is derived by multiplying individual-branch distribution functions for the set of branches with common time sense. The individual-branch distribution functions may also be interpreted as particle field components of the information content of the aether.

This treatment allows the wavefunction both to hold all the information that is ever knowable about a system on this storey and to be the correspondingly real description of reality on this storey. Furthermore the dual time-direction propagation of quantum states ensures that these real descriptions of systems do not fall foul of the relativistic quandaries found in EPR-style experiments.

14.3 Deductions of Eventual Reality

An important characteristic of eventual reality (at least in the flat-space approximation) is provided by the SU(2) and Q_8 dual symmetries of coordinate space and of the aether. These underpinning symmetries are an important prediction of eventual reality. These symmetries are responsible for the nature of eventual-reality's quantitative deductions and they distinguish the theory from other hypothesized descriptions of reality.

These quantitative deductions include:

- the dimensionality of space
- the association of velocity with the rate of change of position
- the mathematical form of the Dirac equation with Dirac matrices
- quantum-mechanical principles
- elementary-particle statistics
- fundamental electrical charges from –3 to +3 units
- three generations of fundamental particles.

Many attempts at forming more fundamental representations of elementary particles, suggest that there should be many other fundamental particles not fitting the pattern derived from the division-ring symmetries. Searches are underway to find such particles using high-energy accelerators. If such fundamental participles (not fitting the dual division-ring symmetries) are ever found, then their discovery will falsify the foundation of eventual reality.

Ultimately perhaps the inclusion of mind as a necessary constituent of the universe will come to be regarded as the greatest achievement of eventual reality. Despite this likelihood, very little detail on aspects of the mind is presented in this document. This current work can only sweep a broad brush over the canvas of reality. Most of the work necessary in all fields is still to be undertaken.

Less formally we have also discussed other things: time-dependent units, the expansion of the universe and lepton-particle masses. It is expected that accurate details of these aspects will dependent upon cosmological analyses involving variable spatial geometry. Note that up to this point in this document, we have assumed space to be Euclidean flat. This approach has greatly simplified matters and has let us derive a basic gravity-free model of quantum mechanics. However, a start has been made with including a variable spatial geometry into eventual reality and this additional work is presented in part 3 of this document.

Notes

[1] At the quantum level, the role of symmetry is most apparent in the field of fundamental particles.

[2] The term continuity is used to describe the smoothness of a field across the quaternion-vector dimensions (eg positional 3-space). Uniformity is used to describe the smoothness of a field in the quaternion-scalar dimensions (eg time).

[3] The second law of thermodynamics requires that the macrostate entropy of a thermally isolated thermodynamic subsystem cannot decrease in time. Thus the entropy always tends to the maximally achievable value for the described system. This maximally achievable value is the stationary value. For example, if we allow mixing of two macrosystems, the maximal condition requires that the total entropy will be equal to the sum of the initial parts or higher.

Epilogue to Part 2

Science is in a continual state of evolution. Sometimes progress in the accumulation of knowledge is slow and smooth, very occasionally change is rapid and disjunct. We cannot tell *a priori* whether we are at a developmental stage where we are ready only for a minor modification to our theories or whether we are in need of a major change to our worldview. However, the premise presumed in this work is that a major reformulation is probably due and hence the full impact of the new theory of eventual reality is worthy of consideration.

We are all aware that our current mainstream theories are deficient in that they do not impose enough restrictions on the natural world for us to predict a unique universe. Unless this defect can be remedied, these current theories will lead us away from a reality that necessarily must be the way that it is actually observed to be, towards acceptance of an unverifiable multitude of potential universes. These different models termed: multi-verses, multi-worlds and parallel universes have limited philosophical justification. Hence, there is value in investigating new and more satisfying worldviews.

There has been much work and some progress in initiating the development of several proposed theories of everything; these theories incorporate a moderate change to the current paradigm. These existing well-established hypotheses (in which the smallest units of matter and or space become extended) are still incomplete but several of these hypotheses have had some recognized successes, for instance, in predicting the entropy of black holes and the necessity for a gravitational field. The further development of these hypotheses is awkward and they are in need of physical guidance as to the nature of reality, this guidance has been lacking. It may be that the difficulties, in completing the formulations of these known hypotheses, arise from the common (Cartesian) mindset that 'everything quantitative' makes reality. It is unknown whether these theories will prove useful in extending the applicability of eventual reality or vice versa.

There have been numerous attempts at a unification of the geometric precepts of general relativity with dynamical theories of matter. These attempts have generally been found to encounter major hazards. Thus it was felt that a reasonable understanding of eventual reality should be developed before this unification is attempted within the conceptual framework of eventual reality. Hence the version of eventual reality presented in part 2 is not a complete theory of the universe.

Ever since Newton proposed his quantitative clockwork universe, scientific philosophy has faced inconsistencies. Modern science has reluctantly admitted that science is a work-in-progress and that (current) science has (unknown) quantitative limitations that make science's predictions accurate but fundamentally imprecise. Eventual reality demonstrates that hypotheses, that displace a quantitative basis of reality in favour of a qualitative basis, re-enable a philosophically sound foundation for our worldview.

Generally when faced with competing alternative explanations that cannot be distinguished experimentally we chose to accept the simplest, most elegant/beautiful explanation. This method of choosing, once known as the principle of parsimony, is now called Ockham's razor. The elegance or beauty of a theory is revealed by the paucity of its introduced concepts and the expansiveness of its scope of applicability. Thus the beauty of a fundamental theory is the measure of how well the theory conforms to the ideal archetypal form of a fundamental theory. The simplicity of a theory is revealed by its ease of use. Clearly it would have been counter productive for Newton to have proposed quantum mechanics instead of his classical laws of motion because it was much easier to use classical mechanics to solve the problems of Newton's time. Hence we see that with these definitions, our measures of beauty are eternal and our measures of simplicity are ephemeral.

Eventual reality is an elegant hypothesis. The hypothesis is seductively simple and extremely powerful with intriguing insights into how the universe is built. Eventual reality is a hypothesis that creates room for consciousness, mathematics, logic and linguistics. The mind is not a property of matter; matter is not a property of mind. Both the mental and material are derived concepts; the foundation concepts are the frenetically linked events that are the raw expression of the universe. Thus the hypothesis has few fundamental concepts and an impressively expansive scope of applicability. Eventual reality requires that there should be laws of nature and it gives the form of these laws. For instance, eventual reality explains both the principles and the rules of quantum mechanics. In addition, eventual reality naturally incorporates the qualitative elements of the universe that quantitative science has not been able to digest. Thus eventual reality survives the cut delivered by Ockham's razor.

Eventual reality was developed as an outcome of a metaphorical leap in the dark as described in a map based upon previous shifts in the scientific paradigm. Eventual reality has many intriguing features and properties. It has yet to be accepted. Time will tell whether the theory is useful in its own right or simply as a demonstration of the power of theories developed around the qualitative. Now, it is difficult to discard powerful scientific principles that have exceeded their unmarked use-by dates. It took a quarter of a century to turn the old quantum mechanics of Planck into the fledgling (non-relativistic/classical field) quantum mechanics of Heisenberg and Schrödinger. We should expect that early versions of hypotheses such as eventual reality, if they were not to be conclusively rejected, would require further development over a reasonable period of years.

Eventual reality has a distinct philosophical basis; nevertheless the theory is still incomplete. However, it conduces to a physical universe much as we experience it. In this derived universe, there is consciousness, there are four physical dimensions to (reduced) space-time, there are fermions and bosons with the known symmetries, there are laws of nature, quantum mechanics and quantum measurement exist as a statistical approximations, etc. These are remarkable achievements for such a juvenile theory.

Beneath the physical world, the immaterial universe (of eventual reality) is describable as a conversation (in a foreign language) distilled from a self-cancelling microscopic cacophony. Intriguingly eventual reality offers ways in which this metaphysical picture may be explored by incorporating the mind as a functional link within a physical experiment.

However, the simple form of the hypothesis (presented in part 2) is inconsistent with the framework of general relativity. Thus it is to be expected that those aspects of the hypothesis that involve universal scales of length or time, details of the early universe, singularities, etc. will require a fully relativistic version of eventual reality. The presented discussions on the expansion of the universe and numerical values and the time evolution of physical constants were given primarily for guidance to elucidate the limiting behaviours and facilitate the development of a more complete theory.

Although the event is the fundamental unit of the universe, for most practical purposes, the universe is viewed as a multi-branched multi-dimensioned (ie pan-dimensional) space filled with the hum of the aethereal langue. The philosophical implications, of constructing coherent reality from qualitative linguistic components, are substantial. Clearly expression of this language can be labelled as thought. However, this general thought is not a derivative of matter; verbalism is seen to be a representation of the fundamental aspects of the universe. Matter is not dreamt up by a thought-producing organ (mind); rather, matter is a statistical derivative of the decaying residue of potential thoughts that comprise the branched hierarchical aethereal structure of the universe.

Finally it also appears that many different schools of philosophers can celebrate. Rationalists can feel satisfied that our universe is a deductive consequence of the principles of eventual reality. Empiricists can proclaim that the entire theory was inferred inductively from experiences (as outlined in part 1). Essentialists may rejoice in the theory's determination of an explicable and ultimate reality. Sceptics can enthusiastically debate the evidential necessity to be sceptical of scepticism. Idealists can delight in the complete perfection and harmony of natural laws that could not be other than they are. Monists may celebrate the construction of everything from a single monadic form. Solipsists can ponder the holism of the aether. Analysts may marvel at the interconnectivity of the universal langue with the material world.

PART 3
Supplementing the Paradigm with Affine Curvature Expressions

Prologue to Part 3

This complete document deals with the structure and appearance of the edifice of *reality*. Part 1 discusses the evolutionary development of scientific worldviews. Part 1 infers a foundation picture revealing a reality based upon an unrestrained repetition of *events*. Part 2 of this document explores this picture and handles the implications of an all-pervasive qualitative *aether* that is responsible for practical experience. In turn, part 3 deals with the large-scale quantitative role of *space* and space's interplay with matter. Thus the greater share of the discussions of part 1 was to do with the development of scientific descriptions of the monadic foundations of reality. The greater share of part 2 was to do with showing how a universe of events could approximate to quantum mechanics on a higher storey of reality. The bulk of part 3 will discuss the substance of the geometric theories such as general relativity and this part will complete our general understanding of the quantitative material aspects of the universe.

Many will agree that general relativity and quantum mechanics are the two most significant and positive intellectual achievements (not limited to the fields of science) of the twentieth century. We have shown how eventual reality relates to quantum mechanics but we have not shown how eventual reality relates to the other major achievement ie general relativity. Now, a framework for guiding the development of new worldviews (largely based upon the development of quantum mechanics) was presented in part 1. This framework was used to develop a new theory of reality. Subsequently, in part 2, a strain of quantum mechanics was derived (from this new theory called eventual reality) for a flat Euclidean space. Part 3 starts by discussing general relativity from the perspective of eventual reality. In this part we shall merge these two approaches (eventual reality and general relativity) and we will confirm that eventual reality fulfils the essential requirements of the legendary 'theory of everything'.

The whole work described in this document has been laid out like an instruction map. Part 1 of this work presented the idea of using a map for exposing new scientific paradigms. The first six steps of the map were followed in part 1 and a new worldview was gained. The crucial seventh and eight steps of the map were completed and described in part 2. These final steps delivered a detailed (but approximate) theory based upon the new worldview exposed in part 1. Part 3 can be variously viewed as a ninth step or even early steps in an alternative journey (the first step is always to decide the destination and summon up the map).

In the immediate aftermath of an appropriate change of paradigm, a new paradigm shift should not be needed. Thus we may need to relinquish some of the belief structure used to construct the map. In part 3 we aim to reformulate the hypothesis of eventual reality, to remove deliberate simplicities and to replace them with other approximations to reveal a new side to the theory. This accomplishment will involve some reinterpretations but the worldview inferred in part 1 and exposed in part 2 will be unchallenged.

When explaining any theory there is a limit on how much the writer should jump forward and draw upon results not yet proven at that point in the argument. In an attempt to render the discussions easily understandable, I may have jumped forward too often for some readers. However, limits were placed upon the degree of such jumping forward and as a result some potentially clarifying discussions were omitted in part 2. In contrast to part 2, part 3 of this document is not so constrained. A comprehensive picture of the theory has already been presented in part 2. Thus part 3 can draw upon knowledge from all sections of the theory to elaborate its arguments.

Chapter 15 : Statuses of the Theories

An introductory form of the theory of eventual reality was presented in part 2 of this document; this introductory form was provisioned with a simple flat structure for space. This flat-space approximation helped to make the theory readily comprehensible and the approximation aided the theory's rapid development. However, in part 3 we will consider how we could expand the theory's foundations by developing a different form of the theory of eventual reality.

Thus in this chapter, we shall start by considering the problematic issues that we have encountered with the theory of eventual reality. These considerations will provide justification for investigating alternative improved mathematical representations of space that incorporate the lessons of general relativity.

15.1 Critique of Eventual Reality

There are several reasons for starting part 3 of this document with a critique of the conclusion achieved in the earlier parts. This document has made many far-reaching claims; it is only fair that the document should contain a word of caution and that we should present an apologetic self-critique. This work is the first presentation[1] of the theory; much more work remains to be undertaken. Primarily a new worldview has been formed and a primitive form of the theory has been developed around the worldview. Thus the normal gradual processes of improvement of the representations, expansion of scope of applicability and investigation of implications begin from here. Furthermore, were it possible to replace eventual reality with an even deeper level theory, then we would also need to identify any worries, inaccuracies or inconsistencies inherent in the theory. Thus we have two good reasons for considering the difficulties that the theory may have.

Let us recall why eventual reality is an unfinished theory and why the version of eventual reality that is presented in part 2, is preliminary. The development took a particular route that concentrated on forming the hierarchy of representations of nature and comprehending the transition to quantum mechanics (and the everyday microscopic worldview). This development has established a clear picture of the nature of the hypothesis but the development has ignored some issues. Primarily this development of eventual reality has ignored the large-scale cosmological aspects. For eventual reality to earn the title of a *theory of everything*, eventual reality would need to include gravity at a fundamental level. However, gravity (and spatial curvature) is absent from the part-2 version of eventual reality. This is a significant omission, one that will be corrected in part 3.

Eventual reality claims to be a parameter-free model. All parameters introduced in this work should be calculable from within the theory. However, at this stage these parameters have not been calculated. A consideration of the masses of leptons has shown how the masses might be derived from the theory. Nevertheless the success of an analytical derivation is not guaranteed by this hand-waving discussion.

Eventual reality requires that measured fundamental constants should display a dependence on time. All attempts to detect such a variation have failed. However, it is premature to predict accurate temporal dependencies for the constants of nature since the principles of general relativity have not been incorporated and a cosmological theory has not been developed.

Eventual reality has been given an *a priori* justification. Though the justification is plausible and rational, on its own, the presented justification appears to be non-exhaustive. There is the taint of an implicit expectation that the universe should be understandable. Whereas this is empirically justifiable, the rational justification for an understandable universe needs strengthening. Furthermore, there could be worries that rational philosophical arguments (as used to justify eventual reality) almost universally have been met with empirical counter arguments which purport to prove an opposing claim. However, the current study as a whole shows that the hypothesis of eventual reality achieves a reconciliation of empirical observation and science with abstract reasoning and mathematics. In this way eventual reality evolves into recognizable science on the higher storeys of reality and adds the realm of mental experience.

A remaining awkwardness in the deduction relates to the intrinsic circularity of logic's justification and use. This loop is created when we apply our higher-storey logic to the foundation storey of reality. Curiously, on the foundation storey, logic and reason are non-existent. The available options form their own discordant apeiron (a term introduced by Anaximander to refer to the boundless origin of everything). In this ground-storey liberated network of events, any and every nonsensical form is permitted without challenge. However, the parading collection of nonsense acts to nullify itself. Our logic emerges out of the coherence of the antagonist lingua franca of the apeiron. Justifiably we apply our logic to the apeiron since our logic delivers an explanation in the only format that we can comprehend. Nevertheless it would be nice to somehow justify the apeiron from within itself.

The unfolding Russian-doll-like regressive and reductionist process of revealing internal material constituent elements (within the material constituent elements of the previous shell) has an ending in hybrid space with a *finite* set of fundamental (fermion and boson) material particles. Beneath this picture, matter is formed from the statistics of aethereal linguistic sentences. Very few details of this universal language have been discussed in this work. Whilst we may access the langue through our thoughts, rigorous scientific style validation of the language's existence and its exploration may prove difficult.

Thus in this section, we have raised quite a few niggly concerns relating to the capabilities of the theory of eventual reality. Some of the scientific issues (eg evaluating various universal parameters) probably simply require further analysis and expansion of interpretations; these issues require no further development of the foundation descriptions of the theory. The other philosophical worries are also unlikely to be serious. Thus for the present we may have to accept these noted philosophical concerns until such time that we are better able to apprehend the apeiron. However, the exclusion of curved space is a serious worry. The absence of curvature provides an escape clause for any experimental falsification of the theory. Hence so that we can be sure of observing any shortcomings in the theory we will need to include the effects of gravity and spatial curvature. This reason is why part 3 is devoted to such gravitational issues.

15.2 Status of Classical General Relativity

General relativity was developed before the birth of quantum mechanics. Thus general relativity is understandably steeped in the ideology of classical mechanics. In the twentieth century, quantum mechanics started with a very simple form and then quantum mechanics underwent many formal upgrades. General relativity has also undergone some development; however, general relativity seems to have been more mature at birth than quantum mechanics was at its birth. Indeed general relativity has been found to be more difficult to test and develop than quantum mechanics and consequently the original form of general relativity has proved reasonably resilient. Furthermore, whereas quantum mechanics is a staple ingredient of every undergraduate physics course; general relativity is not routinely taught in university Physics courses. For this reason, this document will present many of the conceptual ideas and analytical tools of basic general relativity in this chapter. These results are presented in a form that will prove useful in the reformulation (of eventual reality) that we shall make. In contrast, the corresponding background details of quantum mechanics, were deemed to be more widely known and more easily accessible and such details were not given in part 2.

Affine geometry[2] provides an appealing way of introducing general relativity. Initially we shall presume that space is universally flat (eg special relativity) and then subsequently we shall allow space to undergo continuous distortions such that space remains effectively flat on an infinitesimal scale. The elements of this affine geometry are to be tensors. We start the mathematical definitions by associating the mathematical-real numbers with rank-zero tensors (affine scalars). Rank-one tensors (affine vectors) are defined using the affine axioms. The imposed (module-like) affine axioms correspond to the operations of translation and rescaling. These axioms are listed in table 15.1.

Table 15.1

Basic defining axioms for the algebra of an affine-vector system. Vectors are rank-one tensors. Rank-zero tensors (affine scalars) are the mathematical-real numbers.

System			Affine
Affine Vectors	Rules of Addition	Commutative law: $\underline{a}+\underline{b}=\underline{b}+\underline{a}$	x
		Associative law: $\underline{a}+(\underline{b}+\underline{c})=(\underline{a}+\underline{b})+\underline{c}$	x
		Translational law[3]: $\forall\ (\underline{a},\underline{b})$ $\exists\ \underline{c}$ such that $\underline{a}+\underline{c}=\underline{b}$	x
	Rules of Multiplication by Affine Scalar	1st Distributive law: $(\lambda+\mu)\underline{a}=(\lambda\underline{a})+(\mu\underline{a})$	x
		2nd Distributive law: $\lambda(\underline{a}+\underline{b})=\lambda\underline{a}+\lambda\underline{b}$	x
		Associative law: $(\lambda\mu)\underline{a}=\lambda(\mu\underline{a})$	x
	Unitary	Identity element: $1\underline{a}=\underline{a}$	x

Once this affine algebra has been specified, a spatial dimensionality and a spatial metric are posited. The axiom of dimensionality introduces an affine tensor manifold with a fixed number of dimensions. This integer number of dimensions (>1) is supplied but in other respects the number of dimensions is arbitrary. Thus the coordinates of space (rank-one tensors) are introduced as ordered composite sequences of mathematical-reals satisfying the affine algebra of table 15.1.

Note that at this stage there are no rules of multiplication between affine vectors. We shall allow an inner (dot) product (to be discussed in section 15.3) and an outer (Grassmann) product as the only accessible products between affine vectors. The outer product can be used to increment the rank of an affine tensor eg the outer product of two rank-one tensors is a rank-two tensor.

Algebrae can clearly be constructed that are not based upon mathematical-real numbers. Furthermore a more general tensor form would describe the relevant types of number that we have discussed in part 2. Thus general relativity appeals to empirical evidence of the appropriateness of tensor descriptions built using mathematical-real numbers. Recall that eventual reality does not need to make such an empirical appeal: in part 2, it was shown that integers enumerate the size of sets of events; these integers provide a scale from which the number basis for representations of a spatial continuum is derived.

Once the tensorial numerical representation is fixed, we are permitted to consider the physical equivalence of coordinate systems under arbitrary rotations. This equivalence sets an allowable linear homogeneous form for the equations of an affine coordinate transformation.

Let us look forward and briefly mention the steps from this number system to the equation of general relativity. More details of these steps will be provided later. An invariant length formed by an intuitively chosen scalar product (of contravariant and covariant vectors) is introduced to commence the metrification of space. Another requirement is that of a neighbourhood relationship (dubbed affine connexion) that assigns rules for parallel transport of a vector direction. In a flat space, we can carry a vector direction around any closed loop and the directions will line up when they meet again – such a space is called Euclidean affine. However, in a curved space (such as the boundary between the air and land or sea on the Earth), the directions will not generally line up. This misalignment causes problems for mapmakers. The journey of a fictitious crow provides an illustrative example of how a misalignment of direction in the Earth's biosphere can occur. The crow starts at the equator and flies south; the crow knows that his right wing points west. He flies steadily in an apparently straight line over the South Pole and continues without stopping until he returns to the equator. Then the crow turns and he flies around the globe in the direction that had been pointed to by his right wing and eventually the crow meets his stay-at-home brother who tells him that he was most recently flying eastward not westward. Now, we want to be able to reliably assign a consistent direction and therefore be able to construct differential equations. Thus in a curved space we need a formal procedure to provide an objective reference direction at different places.

Continuity of space-time continuum and of the material structures that are impressed upon the continuum, are achieved by applying these extended affine requirements locally ie on an infinitesimal scale. Thus measures of space and time are permitted to vary continuously and smoothly throughout the continuum. These variances from the ideal flat structure are deemed distortions of space-time and the variances are identified with material and gravitational fields.

Matter fields are represented by a rank-two energy-momentum tensor (generally known as the energy tensor). Thus spatial distortions are described by a rank-two tensor expression formed from contractions of the general Riemann curvature tensor. The expression in terms of the curvature is determined by covariance requirements as espoused by a variational principle.

Note that each step in this development of the mathematics underpinning general relativity is geared to producing a continuum space meeting *empirical* expectations. To further expose these issues we shall discuss these steps in more detail.

15.3 Flat-Space Metrification

Physical spaces are symbolic systems that continuously convey from point to neighbouring point a sense of scale and a sense of direction. The spatial metric is a formal representation of the requirements for point-to-point transference of scale and direction (whether temporal, distance-like, energy-like or momentum-like).

Let us now consider how the metric is defined for flat space, firstly by general relativity and secondly by eventual reality. We will explicitly derive or explain those results that we will need for our eventual version of general relativity.

15.3.1 Transport of Measures for Tensors

The metric is introduced as a unique form appearing in a bilinear product. The bilinear product is the analogue of the scalar product of vectors in Euclidean space. Thus it is easiest to introduce the appropriate concepts in a flat (Euclidean or Lorentzian style) space. We define affine metrical vectors, $\underset{\rightarrow}{U}$ and $\underset{\rightarrow}{V}$, in terms of *independent* orthogonal unit directions, $\hat{\eta}_i$, and with component magnitudes u^i and v^i.

$$\underset{\rightarrow}{U} = \sum_i u^i \hat{\eta}_i \qquad \text{and} \qquad \underset{\rightarrow}{V} = \sum_i v^i \hat{\eta}_i \tag{15.1}$$

The invariant bilinear product, Q, *defines* the matrix, g_{ij}, which is called the spatial metric. Although we are using tensor notation, we will *not* adopt the physicist's convention that repeated pairs of (covariant and contravariant) indices implies a summation over the index.

$$Q(\underset{\rightarrow}{U}\underset{\rightarrow}{V}) = \sum_{i,j} u^i g_{ij} v^j \tag{15.2}$$

In Euclidean space this bilinear product is just the product of the lengths of the vectors, $\underset{\rightarrow}{U}$ and $\underset{\rightarrow}{V}$, and the cosine of the angle between them. This value is invariant under changes of coordinate representation (eg an arbitrary rotation of Cartesian axes, a change to the style of coordinates or a change in the unit system). All quantities in this expression (15.2) are commuting mathematical-real numbers. Thus without loss of generality, we may assume that the metric is symmetric under exchange of indices ($g_{ij} = g_{ji}$).

In principle, we should expect that

$$Q(\hat{\eta}_i \hat{\eta}_j) = \delta_{ij} \,. \qquad (15.3)$$

This result defines what we mean, in a metrical space, by the orthogonality and the unitary size of the directional designators, $\hat{\eta}_i$. However, we must not assume the vectors are constructed in a simple designator-free algebra over mathematical-real numbers. We have seen that for division-ring numbers

$$\hat{\eta}_i \hat{\eta}_i = f_i \hat{\eta}_0 \qquad \text{where } f_i = \pm 1 \,. \qquad (15.4)$$

However, this result is not more generally applicable[4]. Nevertheless, to obtain the necessary answer for the empirical signature of flat space, general relativity requires that we drop equation (15.3) and instead assert a result resembling equation (15.4)

$$Q(\hat{\eta}_i \hat{\eta}_j) = f_i \delta_{ij} \,. \qquad (15.5)$$

Thus equation (15.5) has only an empirical justification in general relativity. Now, this bilinear orthogonality (15.5) has diagonalized the metric. Choosing $u^i = \delta^i_k$ and $v^i = \delta^i_l$ we see from equation (15.2) that the metric shares the signs of the set of numbers $\{f_i\}$

$$g_{ij} = f_i \delta_{ij} \,. \qquad (15.6)$$

Next, we consider the quadratic product and how this product may be identified with the square of the length of a vector

$$Q(\underline{U}, \underline{U}) = \sum_{i,j} u^i g_{ij} u^j = \sum_{i,j} u^i (f_i \delta_{ij}) u^j \qquad (15.7)$$

$$= f_0 \left(u^0\right)^2 + f_1 \left(u^1\right)^2 + \ldots f_n \left(u^n\right)^2 . \qquad (15.8)$$

Thus the sequence of numerical signs displayed by the f_i give the signature of space. This result (15.8) is compatible with the scientific observation that Lorentzian space has a signature

$$Q_L(\underline{U}, \underline{U}) = -\left(u^0\right)^2 + \left(u^1\right)^2 \ldots + \left(u^n\right)^2 \text{ for } n = 3 \,. \qquad (15.9)$$

However, if we had not modified the quadratic product (15.5) of the $\hat{\eta}_i$, we would have obtained the Euclidean result (all positive signs) which has a more justifiable vector length, as suggested by Pythagoras' theorem.

From the hindsight of eventual reality, we may appreciate that the difference between the Euclidean quadratic product and the empirical product, $Q_L(\underrightarrow{U}, \underrightarrow{U})$, has a rationale in the reciprocal transformation that turns coordinate space into hybrid space. Explicit expressions for these metrics will be given in the next section.

General relativity uses the metric to define the geometry of space, thus by definition no other information about space is required. Note though in non-Riemann spaces, an additional 1-form may be added to impose a functionally variable scale.

The metric also introduces the concept of covariant and contravariant vector pairs[5]. This pairing is yet another representation of the conjugate pairing explained in chapter 10.

$$Q(\underrightarrow{U}\underrightarrow{V}) = \sum_{i,j} u^i g_{ij} v^j = \sum_i u^i v_i = \sum_j u_j v^j \tag{15.10}$$

Here the metric is used to describe the operation that transforms tensors between their conjugate-pair forms as

$$u_i = \sum_j g_{ij} u^j \text{ and } w_j^i = \sum_k w^{ik} g_{kj} \tag{15.11}$$

with matching equations

$$u^i = \sum_j g^{ij} u_j \text{ and } w_j^i = \sum_k g^{ik} w_{kj} . \tag{15.12}$$

Note that these equations illustrate how the metric transforms higher-rank tensors. Hence

$$w_{ij} = \sum_l g_{il} w_j^l = \sum_{lk} g_{il} g^{lk} w_{kj} . \tag{15.13}$$

Since the second-rank tensor, w_{ij}, is arbitrary we must have

$$\sum_l g_{il} g^{lk} = \delta_i^k . \tag{15.14}$$

We note that δ_i^j is the tensorial Kronecker delta function with invariant unit normalization. Thus an invariant derivative (which is denoted by the semi-colon and will be defined in section 15.6) of the tensorial Kronecker delta function vanishes

$$\delta_{i;k}^j = \delta_{i,k}^j = 0 . \tag{15.15}$$

Note that we do not write δ_{ij} for the fully covariant version of δ_i^j, we prefer to write this covariant expression as g_{ij}.

15.3.2 Flat Space Interpreted by Eventual Reality

Now, in eventual reality, the coordinate space (the foundation space) is derived from the apeiron with limited empirical guidance. The discrete counting of events provides an *a priori* measure of separation. No external form of mathematics is imposed and, for flat space, the appropriate algebra is revealed as that of a pair of division rings over the mathematical-real numbers. In part 2, this algebra was shown to generate the qualitative storeys of reality. Before we consider modifying the eventual reality of part 2 by including spatial distortion, we will consider how eventual reality can justify some of the empiricisms within the theory of general relativity.

Let us consider a general coordinate and consider its algebraic conjugate (ie mathematical-complex conjugate *and* quaternion conjugate) denoted by a dagger†

$$\underline{\underline{\Xi}} = \sum_i \xi_i \hat{\eta}_i \qquad \text{and} \qquad \underline{\underline{\Xi}}^\dagger = \sum_i f_i \xi_i \hat{\eta}_i \,. \tag{15.16}$$

The product of these coordinates gives the invariant scalar product (equal to the square of the length)

$$\underline{\underline{\Xi}}^\dagger \underline{\underline{\Xi}} = \sum_i f_i \xi_i \xi_i \hat{\eta}_i \hat{\eta}_i + \sum_{\substack{i,j \\ i \neq j}} f_i \xi_i \xi_j \hat{\eta}_i \hat{\eta}_j \,. \tag{15.17}$$

The first term on the right hand side of equation (15.17) is the rational Euclidean result; the second term is a torsional term, which vanishes by symmetry. Perhaps surprisingly the algebra of a mathematical-complex quaternion number form does not necessarily imply that the torsional term vanishes by cancellation. For instance, representing a mathematical-complex quaternion as a pair of a scalar and a vector (both mathematical-complex)

$$\left(z^*, -\underline{v}^*\right)\left(z, \underline{v}\right) = \left(z^* z + \underline{v}^* \cdot \underline{v}, z^* \underline{v} - z \underline{v}^* - \underline{v}^* \times \underline{v}\right). \tag{15.18}$$

The omission of the torsional term in (15.17) only applies when the numbers factorize into a quaternion for which we have

$$\left(a, -\underline{w}\right)\left(a, \underline{w}\right) = \left(a^2 + \underline{w} \cdot \underline{w}, a\underline{w} - a\underline{w} - \underline{w} \times \underline{w}\right) = \left(a^2 + \underline{w} \cdot \underline{w}, 0\right) \tag{15.19}$$

and a mathematical-complex number for which we have

$$(x, -y)(x, y) = (x^2 + y^2, xy - yx) = (x^2 + y^2, 0) \,. \tag{15.20}$$

Note that this algebraic multiplication (15.18) is a mathematical-complex quaternion product; however, this product is equivalent to a quadratic scalar/inner product as defined by equation (15.7). The product predicts a Euclidean signature for coordinate space

$$\underline{\underline{\Xi}}^{\dagger}\,\underline{\underline{\Xi}} = \sum_i \xi_i^2 \,. \qquad (15.21)$$

Now, this is an algebraic product; it is not the partially fabricated scalar product, Q. Thus we can fully justify using the quadratic product to depict the square of an invariant length. The rules of quaternion algebra and of mathematical-complex algebra deliver a scalar (ie identity element of the designator group) term from the products of such symmetric numbers. In both algebrae (tensor and mathematical-complex quaternion), the square of the invariant length is defined by multiplying a number by its paired conjugate. Thus the quadratic product allows us to transport both a length measure and a sense of direction from place to place *in a flat space*. To confirm the technique of parallel transport of direction, let us renormalize two numbers to have unit magnitude. Then these unit-magnitude numbers represent parallel directions if their scalar product is unity.

However, the quadratic product (15.2) is only adequate for multiplying null tensors and for multiplying tensors differing only in magnitude. All other detail has been expunged from the analysis. Thus tensor algebra misses the character of quantum matter.

15.4 The Metrics of Flat-Space Eventual Reality

Let us now be more explicit about the mathematics that corresponds to the flat-space versions of eventual reality. Conventionally the coordinate-space metric on each branch is

$$_{cs|br}g_{flat} = g_{Q_s} \otimes g_{C_s} \qquad (15.22)$$

where the quaternion metric is

$$g_{Q_s} = \begin{bmatrix} (+)_P & 0 & 0 & 0 \\ 0 & (+)_P & 0 & 0 \\ 0 & 0 & (+)_P & 0 \\ 0 & 0 & 0 & (+)_P \end{bmatrix}. \qquad (15.23)$$

The quaternion metric is combined with the mathematical-complex metric

$$g_{C_s} = \begin{bmatrix} (+)_T & 0 \\ 0 & (+)_T \end{bmatrix}. \qquad (15.24)$$

Thus with the axis-ordering convention explained in table 10.1, the metric for each branch of flat coordinate space is uninteresting despite being founded upon a more interesting algebra

$$
{}_{cs|br}g_{flat} = (+)_P(+)_T \begin{bmatrix} 1 & 0 & 0 & 0 & 0 & 0 & 0 & 0 \\ 0 & 1 & 0 & 0 & 0 & 0 & 0 & 0 \\ 0 & 0 & 1 & 0 & 0 & 0 & 0 & 0 \\ 0 & 0 & 0 & 1 & 0 & 0 & 0 & 0 \\ 0 & 0 & 0 & 0 & 1 & 0 & 0 & 0 \\ 0 & 0 & 0 & 0 & 0 & 1 & 0 & 0 \\ 0 & 0 & 0 & 0 & 0 & 0 & 1 & 0 \\ 0 & 0 & 0 & 0 & 0 & 0 & 0 & 1 \end{bmatrix}.
$$

There are several differences between this metric and a Lorentzian metric. The differences include the metric's dimensionality of 8, the presence of two incompatible sets of numeric ± signs and inevitably the signature of space.

To describe the general pan-dimensional space, we need to repeat the quaternion and mathematical-complex symmetries for the branches. The quaternion-branch metric is

$$
g_{Q_b} = \begin{bmatrix} (+)_C & 0 & 0 & 0 \\ 0 & (+)_C & 0 & 0 \\ 0 & 0 & (+)_C & 0 \\ 0 & 0 & 0 & (+)_C \end{bmatrix}. \qquad (15.25)
$$

The complex-branch metric is

$$
g_{C_b} = \begin{bmatrix} (+)_B & 0 \\ 0 & (+)_B \end{bmatrix}. \qquad (15.26)
$$

Recall that commonality of space across the branches associates the action of applying some of these mathematical signs (see section 10.9.3).

Now, the Fourier transformation can be considered to reveal the phenomenon of matter and to switch the $(-)_T$ sign of the metric's quaternion-vector components when we reorder the coordinates. Hence the hybrid space metric per particle per branch is

$$
{}_{hs\not{p}r}g_{flat} = (+)_P \left[\begin{array}{cccc:cccc}
(+)_T & 0 & 0 & 0 & 0 & 0 & 0 & 0 \\
0 & (-)_T & 0 & 0 & 0 & 0 & 0 & 0 \\
0 & 0 & (-)_T & 0 & 0 & 0 & 0 & 0 \\
0 & 0 & 0 & (-)_T & 0 & 0 & 0 & 0 \\
\hdashline
0 & 0 & 0 & 0 & (+)_T & 0 & 0 & 0 \\
0 & 0 & 0 & 0 & 0 & (-)_T & 0 & 0 \\
0 & 0 & 0 & 0 & 0 & 0 & (-)_T & 0 \\
0 & 0 & 0 & 0 & 0 & 0 & 0 & (-)_T
\end{array}\right].
$$

Despite the $(-)_T$ sign switch, the mathematical-complex symmetry is retained unchanged in this equation:

$$
{}_{hs\not{p}r}g_{flat} = g_{Q_h} \otimes g_{C_s}. \qquad (15.27)
$$

However, the quaternion part of the hybrid-space metric has become

$$
g_{Q_h} = \begin{bmatrix}
(+)_P(+)_T & 0 & 0 & 0 \\
0 & (+)_P(-)_T & 0 & 0 \\
0 & 0 & (+)_P(-)_T & 0 \\
0 & 0 & 0 & (+)_P(-)_T
\end{bmatrix}. \qquad (15.28)
$$

Thus in this 4-dimensional hybrid space (which we shall call a reduced space), the quadratic product gives

$$
Q(\underset{\to}{U},\underset{\to}{U}) = \sum_{i,j} g_{ij}u^i u^j = (-)_T\left[(-)_T(u^0)^2 + (u^1)^2 + (u^2)^2 + (u^3)^2\right] \qquad (15.29)
$$

which is in agreement with the empirical Lorentzian signature of flat space[6]. Furthermore, the equation displays only one mathematical sign that sign associated with time reversal; the quaternion sign is redundant in the metric.

Let me remark that just two additional steps will take us from this presented analysis to that of general relativity. These two steps are to be discussed fully later. In the first step, we acknowledge that the scale-lengths and directions may vary from point-to-point. However, we require changes in scale-lengths and directions to be vary continuously. Hence, deviations from the Lorentzian result must be either first or a higher order ie we preserve the flat-space symmetries on the infinitesimal scale of continuum space.

Secondly we acknowledge that flat space is gravity free and thus flat space is matter free. The curvature tensor is a suitable invariant to describe distortions of space and hence gravitational fields due to matter. Hence, equation invariance (for all physical laws) will allow us to deduce the Einstein-Ricci equation of general relativity.

15.5 Understanding Reduced Spaces

In coordinate space, the coordinate and hence the metric are factorable into two interwoven subspaces. We perform a transformation on the mathematical-complex subspace to generate hybrid space. Hence we are able to work with two interrelated (non-independent) reduced four-dimensional spaces. In these reduced spaces let us directly observe the flat-space relativistic correspondences between our mathematical-complex quaternion formulation of eventual reality (part 2) and the tensorial metric representation.

For notational convenience we will represent the time coordinate in distance units ie we set

$$X_0 = cT \,. \qquad (15.30)$$

We construct physically meaningful reduced (four-dimensional) spaces by re-associating the eight hybrid-space coordinates such that we re-link time with distance and we re-link the conjugate velocity terms. Thus we obtain the two reduced 4-dimensional spaces outlined in table 15.2.

Note that the reorganization of coordinates is achieved by introducing the mathematical symbol, i, which is mathematically equivalent to the identifier, $\hat{\imath}$, except that we are required to treat i as belonging to the argument of the number rather than to the designator block. Thus we treat i as though it were the square root of minus one. In keeping with this denotation, the sign $(-)_T$ accompanying i will simply be written without subscript. No other unattached negative sign may be written without a subscript.

The quaternion designators contribute to an all-positive signature in the sign $(+)_P$; however, the complex-imaginary nature of the quaternion-vector component arguments, $-iX_\lambda$ or iW_λ, delivers a metric signature of $(+,-,-,-)$ as expected in a Lorentzian space-time continuum.

Thus in the reduced spaces we independently write an infinitesimal invariant time-interval

$$ds^2 = dX_0^2 - dX_1^2 - dX_2^2 - dX_3^2 \qquad (15.31)$$

and velocity-interval

$$dv^2 = dW_0^2 - dW_1^2 - dW_2^2 - dW_3^2 \,. \qquad (15.32)$$

Table 15.2

Comparison between the notational conventions applying to the spatial coordinates representing reduced 4-space and hybrid space.

	Designator	Hybrid Space	Reduced 4-Space	Reduced 4-Space
0	$\hat{r}\hat{\gamma}_0\hat{e}_V$	q_0, w_0 or $\varepsilon / m_0 c^2$		W_0
1	$\hat{r}\hat{\gamma}_1\hat{e}_F^{(j)}$	q_1 or HX_1 / c		iW_1
2	$\hat{r}\hat{\gamma}_2\hat{e}_F^{(j)}$	q_2 or HX_2 / c		iW_2
3	$\hat{r}\hat{\gamma}_3\hat{e}_F^{(j)}$	q_3 or HX_3 / c		iW_3
4	$\hat{i}\hat{\gamma}_0\hat{e}_V$	q_4 or HT	$X_0 = cq_4 / H$	
5	$\hat{i}\hat{\gamma}_1\hat{e}_F^{(j)}$	q_5, W_1 or $P_1 / m_0 c$	$-iX_1 = -icq_1 / H$	
6	$\hat{i}\hat{\gamma}_2\hat{e}_F^{(j)}$	q_6, W_2 or $P_2 / m_0 c$	$-iX_2 = -icq_2 / H$	
7	$\hat{i}\hat{\gamma}_3\hat{e}_F^{(j)}$	q_7, W_3 or $P_3 / m_0 c$	$-iX_3 = -icq_3 / H$	

We let

$$V^2 = \left(\frac{dX_1}{dX_0}\right)^2 + \left(\frac{dX_2}{dX_0}\right)^2 + \left(\frac{dX_3}{dX_0}\right)^2 . \qquad (15.33)$$

Here V is the speed (with respect to that of light). Then attributing the sign of dX_0 to ds, we observe that clock rates are set within an inertial reference frame as

$$ds = dX_0 \sqrt{1 - V^2} . \qquad (15.34)$$

Next, we will confirm that this predicted clock-rate result correctly reflects the time dilation required (by special relativity) across inertial frames. We shall illustrate the result by discussing this equation (15.34) with reference to a version of the twin paradox.

The path of the twins is sketched in figure 15.1 from the point of view of the stay-at-home twin. The interval length of each path is an invariant. Thus both twins will agree on these values for the interval length along the two paths. Furthermore, in the stay-at-home twin's *inertial* frame, the two meetings are separated by the fixed time interval ΔX_0.

Firstly we will apply equation (15.34) for both journeys as seen by the stay-put twin.

$$\Delta s_1 = \Delta X_0 \Big)_{stay-put} \quad \text{and} \quad \Delta s_2 = \sqrt{1 - V^2}\, \Delta X_0 \Big)_{stay-put} \qquad (15.35)$$

Textural subscripts are used to denote which of the twins, is measuring non-invariants.

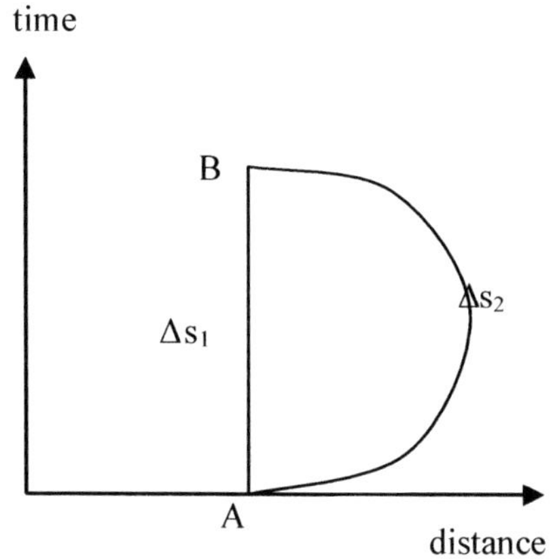

Figure 15.1

Sketch illustrating the twin paradox in the reference frame of the stay-at-home twin who lives in an inertial reference frame whilst her adventurous brother travels the universe at a uniform relativistic speed (apart from take-offs and landings) and returns home.

In the adventurous twin's reference frame, we may only write

$$\Delta s_2 = \Delta X_0\big)_{adventurer}. \qquad (15.36)$$

Combining these equations we can relate the time duration experienced by each twin between their meetings

$$\Delta X_0\big)_{adventurer} = \sqrt{1-V^2}\,\Delta X_0\big)_{stay-put} = \sqrt{1-V^2}\,\Delta s_1. \qquad (15.37)$$

Hence for the adventurer, the elapse time equals the proper time interval of his journey and the elapse time is suitably shorter than the proper time interval for the journey of his more sedentary sister. Indeed his sister takes the longest possible time to travel between the two meetings (A and B).

Now, let us consider the complementary equation (15.32). Equations involving infinitesimals of hybrid coordinates can be factored into equivalent mathematical-real and mathematical-imaginary equations. Inertial frames allow us to extend the infinitesimal factorization to non-differentials. Hence the coefficients in equations (15.31) and (15.32) can only differ by a numerical factor and

$$\frac{dW_\lambda}{dW_0} = \frac{dP_\lambda c}{dE} = \frac{dX_\lambda}{dX_0} \quad \text{for} \quad \lambda = 1\ldots3. \qquad (15.38)$$

Thus

$$V^2 = \left(\frac{dW_1}{dW_0}\right)^2 + \left(\frac{dW_2}{dW_0}\right)^2 + \left(\frac{dW_3}{dW_0}\right)^2 . \qquad (15.39)$$

From which we deduce that the invariant length of the velocity four-vector in an inertial reference frame is described by

$$1 = W_0\sqrt{1-V^2} \qquad (15.40)$$

or

$$m_0c^2 = E\sqrt{1-V^2} . \qquad (15.41)$$

Thus all the hybrid coordinates are seen to relativistically scale in a similar manner.

Table 15.3

The representation of hybrid-space coordinates in flat space illustrates the correspondence between eventual reality and special relativity.

Reduced Space	Contravariant	Covariant
Position	$(X_0, -iX_1, -iX_2, -iX_3)$	(X_0, iX_1, iX_2, iX_3)
Velocity	(W_0, iW_1, iW_2, iW_3)	$(W_0, -iW_1, -iW_2, -iW_3)$

The tensor description is built from contravariant and covariant tensors. Thus to complete the association of the metric-based results with those results of part 2, we use the metric to convert between contravariant and covariant representations of the coordinates. These flat-space associations are Hermitian and they are shown in the following table 15.3. By standard convention we shall only work with contravariant reduced-hybrid-space coordinates for both 4-position and 4-velocity.

15.6 Connexions

Formulae for many spatial properties are modified when we allow space to become distorted. However, on the infinitesimal scale we will assume that space is asymptotically flat. This assumption (of general relativity) is just like postulating that a curve asymptotes to a tangent. The required smoothness of space has already been explained in eventual reality as arising from large-scale averaging (section 10.1).

In flat space, we can carry both a scale length for distance and a sense of direction around in space without the prospect of a disagreement. In curved space, such transported standards are path dependent and thus we need to impose some external regulation so that we can sensibly compare viewpoints. On an infinitesimal scale, this regulated dictate will assert how vectors at neighbouring points can be compared. In a tensor manifold this relationship between neighbouring points is called an affine connexion.

In flat space, vector gradient (grad) and vector divergence (div) operations provide appropriately ranked tensors ie these vector functions represent coordinate independent algebraic operations. In curved space, we seek the analogues of these operators ie we seek invariant differential operators that generate further tensors of higher rank. We can visualize how these differential analogues can be constructed: locally we map space onto the flat tangent space of the location; we apply the corresponding differential vector operator (grad or div) in flat tangent space; then we map the new tensor back onto curved space.

Ordinary derivatives will be indicated by a comma subscript, tensorial covariant derivatives will be indicated by a semi-colon subscript. With linear homogeneous affine transformations, the tensor covariant derivatives of rank-zero and of rank-one tensors must look like

$$A_{;i} = A_{,i} \qquad (15.42)$$

$$B_{k;i} = B_{k,i} - \sum_{l} \Gamma^{l}{}_{ki} B_{l} \cdot \qquad (15.43)$$

A rank-zero tensor is undistorted by the mapping processes. However, a rank-one tensor will acquire a linearized correction to the derivative. The coefficients, $\Gamma^{l}{}_{ki}$, are independent of the components B_l and they may only depend upon the geometry of space (ie the fundamental metric tensor). These coefficients are called the affine connexion or affinity. We will describe how formulae for the covariant derivative of higher-ranked tensors can be obtained from the above equations.

Now, a special example of a covariant rank-one tensor is $A_{;i}$. For this tensor

$$A_{;ki} = A_{,ki} - \sum_{l} \Gamma^{l}{}_{ki} A_{;l} \cdot \qquad (15.44)$$

With our mathematical-real number representation for spatial coordinates, we can interchange the order of differentiation in the tangent-mapped flat space (and hence we can interchange the covariant derivatives of a rank-zero tensor in curved space). The affinity is not dependent upon the tensor suggesting that the affinity may be symmetry. However, not all rank-one tensors are of the form $A_{;i}$; nevertheless we shall assert that the affinity has symmetry[7]

$$\Gamma^{l}{}_{ki} = \Gamma^{l}{}_{ik}. \qquad (15.45)$$

Now, recall that a general rank-two tensor is the sum of tensors of the (outer product) form $B_i\beta_j$. The reversible mapping requires that the tensor covariant derivatives of such products satisfy

$$(B_i\beta_j)_{;k} = B_{i;k}\,\beta_j + B_i\,\beta_{j;k}\,. \qquad (15.46)$$

Hence, a rank-two tensor transforms as

$$C_{ij;k} = C_{ij,k} - \sum_n (\Gamma^n{}_{ik} C_{nj} + \Gamma^n{}_{jk} C_{in})\,. \qquad (15.47)$$

Likewise, we generate expressions for the tensorial covariant derivatives of even higher-ranked covariant tensors.

Next, we shall show how we can obtain the corresponding equations for covariant derivatives of *contravariant* tensors. Note that from equation (15.42), the comma and semi-colon differentials of a scalar invariant are equivalent

$$\sum_k (B^k\beta_k)_{;i} = \sum_k (B^k\beta_k)_{,i} \qquad (15.48)$$

and hence[8]

$$B^k{}_{;i} = B^k{}_{,i} + \sum_l \Gamma^k{}_{li} B^l\,. \qquad (15.49)$$

Thus we obtain the covariant derivative of the lowest order mixed (covariant and contravariant) tensor

$$C^j{}_{k;i} = C^j{}_{k,i} + \sum_n \left(C^n{}_k \Gamma^j{}_{ni} - C^j{}_n \Gamma^n{}_{ki}\right). \qquad (15.50)$$

As previously we may generalize the result to higher-order mixed tensors.

Now, we devise an expression for the affinity (affine connexion) by recognizing the unique characteristics of our two reversibly mapped spaces. The special feature of an asymptotically flat Lorentzian tangential space is that the gradient of the metric is zero. The corresponding special feature of corresponding curved space is that the tensor covariant differential of the metric is zero.

$$g_{ik;l} = 0 \qquad (15.51)$$

Interestingly this result provides another route to the extensions of these expressions to generate the covariant tensor derivatives of contravariant and mixed tensors. Furthermore, this result implicitly fixes a symmetric affinity. Consequently in curved space we have

$$g_{ik;l} = g_{ik,l} - \sum_m \left(g_{mk}\Gamma^m{}_{il} + g_{im}\Gamma^m{}_{kl}\right) = 0\,. \qquad (15.52)$$

Making use of the symmetry (15.45) of the affinity, we may reorganize this equation (15.52) and obtain the set of explicit equations[9] for the affinity

$$\Gamma^l{}_{ik} = \left\{ \begin{matrix} l \\ ik \end{matrix} \right\}_{Christoffel} = \tfrac{1}{2} \sum_m g^{lm} \left(\frac{\partial g_{im}}{\partial q_k} + \frac{\partial g_{mk}}{\partial q_i} - \frac{\partial g_{ki}}{\partial q_m} \right) \cdot \qquad (15.53)$$

Indeed all the topological aspects of spatial deviations from a Euclidean-affine flatness are represented by the affinity. Now, the metric's comma derivative $g_{ik,l}$ is not a tensor. Hence, we observe from equation (15.52) that the affinity $\Gamma^l{}_{ik}$ is not a tensor. However, we can cancel the non-tensorial part of the comma derivative contributions and construct a tensor from (curl-like) tensorial covariant derivatives of $\Gamma^l{}_{ik}$. This procedure gives a useful general expression, the rank-four Riemann curvature tensor, for the geometric distortion of a continuum space

$$R^m{}_{ikj} = \Gamma^m{}_{ij;k} - \Gamma^m{}_{ik;j} \qquad (15.54)$$

where

$$\Gamma^m{}_{ij;k} = \Gamma^m{}_{ij,k} - \sum_n (\Gamma^n{}_{ik}\Gamma^m{}_{nj} + \Gamma^n{}_{jk}\Gamma^m{}_{in}) \qquad (15.55)$$

and

$$\Gamma^m{}_{ik;j} = \Gamma^m{}_{ik,j} - \sum_n (\Gamma^n{}_{ij}\Gamma^m{}_{nk} + \Gamma^n{}_{jk}\Gamma^m{}_{in}) \cdot \qquad (15.56)$$

Thus the rank-four Riemann-curvature tensor is given by

$$R^m{}_{ikj} = \Gamma^m{}_{ij,k} - \Gamma^m{}_{ik,j} + \sum_n (\Gamma^n{}_{ij}\Gamma^m{}_{nk} - \Gamma^n{}_{ik}\Gamma^m{}_{nj}) \qquad (15.57)$$

which displays symmetry

$$R^m{}_{ikj} = -R^m{}_{ijk} \cdot \qquad (15.58)$$

The significance of this curvature tensor is revealed by consideration of the second-order covariant derivatives of a general rank-one tensor. The first-order covariant derivative subtracts out the distortions of space; however, a corresponding subtraction does not hold for the second-order derivative. Thus when we consider the second covariant derivative of a rank-one tensor, B_i, we find

$$\begin{aligned} B_{i;jk} = B_{i,jk} &- \sum_l \left(\Gamma^l{}_{ij} B_{l,k} + \Gamma^l{}_{ik} B_{l,j} + \Gamma^l{}_{jk} B_{i,l} \right) \\ &+ \sum_{lm} \left(-\Gamma^m{}_{ij,k} + \Gamma^l{}_{ik}\Gamma^m{}_{lj} + \Gamma^l{}_{jk}\Gamma^m{}_{li} \right) B_m \cdot \end{aligned} \qquad (15.59)$$

Thus for any rank-one tensor, B_i, the particular second-order derivative expression

$$B_{i;kj} - B_{i;jk} = \sum_m R^m{}_{ikj} B_m \qquad (15.60)$$

reflects the contribution from the curvature of background space.

Two contractions of this tensor (15.57) will be of particular interest. The first contraction delivers the rank-two curvature tensor, this contraction is also known as the Einstein-Ricci tensor

$$R_{ij} = \sum_k R^k{}_{ikj} \cdot \tag{15.61}$$

Hence, this contraction can be written as

$$R_{ij} = \sum_k \left(\Gamma^k{}_{ij,k} - \Gamma^k{}_{ik,j} + \sum_n \left(\Gamma^n{}_{ij} \Gamma^k{}_{nk} - \Gamma^n{}_{ik} \Gamma^k{}_{nj} \right) \right) \cdot \tag{15.62}$$

An alternative contraction

$$S_{ij} = \sum_k R^k{}_{ijk} = \sum_k \left(\Gamma^k{}_{kj,i} - \Gamma^k{}_{ki,j} \right) \tag{15.63}$$

yields no further information since S_{ij} is the skew part of R_{ij}.

The other contraction of interest is the scalar of curvature or Ricci scalar formed by contracting the Einstein-Ricci tensor of equation (15.61)

$$R = \sum_{ij} g^{ij} R_{ij} = \sum_j R^j{}_j \cdot \tag{15.64}$$

15.7 Tensor Characteristics

Finally we shall discuss two useful tensorial parameters that can help us to interpret equations involving tensors. The principal parameter is called the covariant weight[10] of a tensor; the covariant weight indicates the degree of covariance possessed by a tensor. Scalar invariants are attributed a zero covariant weight; the elements of the spatial metric, g_{ij}, are given a unit covariant weight. Relevant examples of the covariant weight of tensors are given in table 15.4.

Unit dimensional analyses are not readily applicable when we work with dimensionless coordinates. Nevertheless sometimes it can be useful to examine how tensorial expressions scale with explicit powers of the coordinate dimensions. Examples of values of this coordinate power are presented in table 15.5. Since in practical measurement we do not use an event-based scale, this expression for the coordinate power can help estimate order of magnitude scaling values for various terms in expressions with different coordinate powers.

Table 15.4

Examples of values of covariant weight in a tensorial space of n-dimensions. In keeping with tradition, infinitesimal spatial coordinates transform as contravariant or covariant according to whether they appear in the nominator or denominator of an expression.

Covariant Weight	Metric	Scalar invariants	Spatial coordinates	Affinity	Curvature
-n/2	$\sqrt{\lVert g^{ij} \rVert}$		$d\tilde{q}$		
-1	g^{ij}				
-½			dq_i		
0	g_i^j	ds, s, $\sum_j A^j A_j$			R
½				$\Gamma^k{}_{ij}$	
1	g_{ij}			$\Gamma^m{}_{ij,k}$	$R^m{}_{ijk}$, R_{ij}
n/2	$\sqrt{\lVert g_{ij} \rVert}$				

Table 15.5

Examples of the way different tensor quantities dimensionally scale with spatial variables in a tensorial space of n-dimensions.

Coordinate Power	Metric	Scalar invariants	Spatial coordinates	Affinity	Curvature
-2				$\Gamma^m{}_{ij,k}$	$R^m{}_{ijk}$, R_{ij}, R
-1				$\Gamma^k{}_{ij}$	
0	g^{ij}, g_i^j, g_{ij}				
1		ds, s	q_i, dq_i		
n			$d\tilde{q}$		

Chapter 15

Notes

[1] The author is concerned with presenting a general understandable account of the basic ideas of the theory. Hence the emphasis is more on how the theory will be received and judged than with what useful new science can be predicted.
[2] Affine geometry is discussed in Weyl (1952) pp17-33.
[3] In shorthand notation, $\forall$ abbreviates *for all* and $\exists$ stands for *there exists*.
[4] A general number system does not satisfy this result since one cannot renormalize a number system with property $\hat{\eta}_i \hat{\eta}_i = 0$.
[5] Subscripts denote covariance; superscripts denote contravariance.
[6] The signature of a flat space was discussed in section 9.3.1.
[7] It is not uncommon for some workers to propose a non-symmetric affinity.
[8] We use the product rule and the generality of tensor β_k.
[9] The algebra is straightforward eg see Schrödinger (1950) p66.
[10] See Weyl (1952) p127 for a discussion on covariant weight.

Part 3

Chapter 16 : Exploring Maps

The map (presented in part 1) has proved a valuable tool by providing a convenient methodology for uncovering a simple form of the theory of eventual reality. As we have noted, the belief structure of the map cannot directly help us improve this simple form of the theory: the map is designed to engineer paradigm shifts. Nevertheless, in our current circumstances, the map can facilitate a reformulation of eventual reality by another route. Observe that in parts 1 and 2, we inadvertently looked for a replacement for the small-scale quantum-mechanical worldview (rather than a replacement for the large-scale general-relativistic worldview) and we devised a form of eventual reality that did indeed explain quantum mechanics. If instead we concentrate upon a replacement to the worldview of general relativity, we might expect to devise a different form of the theory of eventual reality, a form that can explain how general relativity fits within the framework of scientific theories. Knowledge of the earlier (part 2) development of eventual reality will influence this second development and this knowledge is not to be resisted. The cross contamination of concepts will enable the correspondences of the different forms of eventual reality to be more easily recognized.

Before we present this route, it is useful to further confirm the reliability of the map by showing how the map that we have already presented can be used to categorize the steps in the development of general relativity. We shall describe these steps from the prevenient paradigm of classical mechanics with special relativity.

16.1 Classical Mechanics to General Relativity

Let us consider how a fictional development of general reality can be fitted within the framework of the map. The ordering of the cartographical steps should not be regarded as historically relevant. The aim of this section is to show that general relativity could have been guided by recourse to the *science by numbers* strategy delineated in the map. Thus we shall briefly go through the first six steps of the map to develop the overall picture but not the steps involving the formulation of general relativity. The consummative steps, steps 7 and 8, would deliver the known formalism theory of general relativity; thus it is unnecessary to repeat them.

Chapter 16

16.1.1 Step 1: Map

The map outlined in part 1 of this work is both recent and pertinent. Thus we shall adopt the map without an update. The inherent belief structure of the map rests in the so-called theories of similar form, historical revisionism, difficult change and fresh eyes. The successful execution of the next steps of the map will confirm that we can conceptualize the development of the foundation picture of general relativity in these cartographical terms.

16.1.2 Step 2: Dark Clouds

Step 2 asks us to identify what is currently problematic with the accepted (classical) scientific approach and what extra missing features should we accept to generate rational explanations of these dark-cloud problems. Now, there are three suspicious difficulties with classical mechanics. Firstly classical mechanics does not explain the mysterious equality of inertial and gravitational masses; secondly classical mechanics does not have an *a priori* explanation of the Newtonian paradigm or of Newton's laws of motion. Thirdly classical mechanics has a quantitative problem with the precession of the perihelion of the orbit of Mercury. Here we are ignoring all quantum-related issues with classical mechanics.

General relativity took the first problem to be the major dark-cloud issue. General relativity proposed that the solution was to add a requirement for a formal equivalence of gravitation and acceleration. Note that special relativity has already corrected those difficulties of classical Newtonian mechanics, associated with an absolute space (an absolute space clashed with results of classical electromagnetism and with experiment). The second mentioned issue is an unspecific call for an improvement to the worldview. Every improvement to the scientific worldview will provide explanations that are more expansive. Thus such worries only become dark clouds when one attempts to formulate an ultimate theory of everything. Undoubtedly the third issue turned out to be more pertinent than it looked historically. However, it is not possible to demonstrate that this celestial observation must be contrary to classical physics – for instance, no one has yet directly measured the internal organization of the Sun. There are too many unlikely but possible explanations, for this effect to be regarded as a clear deficiency in the classical formalism. Thus this problem identified for a sole planetary orbit cannot easily be attributed to a necessary conceptual omission and hence this disagreement is not chosen as a candidate for the dark-cloud topic.

Now, what we add to explain this phenomenon of equivalent masses is a universal principle of equivalence in the form of a geometrical correspondence of gravitation and acceleration.

16.1.3 Step 3: Master Principle

The master principle is to be one of the principal planks used in the construction of the theory. The master principle is required to be a generalization of the precept introduced to explain an identified dark cloud. The dark-cloud addition, selected in step 2, is the asserted equivalence of gravitation and acceleration. Einstein generalized this formal equivalence of gravitation and acceleration into the master principle that states that no coordinate frame has any *a priori* privilege over any other coordinate frame. Thus the laws of nature (but not the descriptions of natural processes) must be precisely the same for all observers (in any state of motion or experiencing any system of forces); furthermore the laws of nature are expressible in a form that is independent of any particular coordinate representation. Note that this general invariance of the natural laws does not stop some reference frames from being more special than other frames. Frames can have a special status by virtue of the perceived organization of matter and fields in that frame. It is the rules that matter obeys that cannot be special.

16.1.4 Step 4: Duality

Relevant dualities associate seemingly opposing and competing concepts that resist being replaced by a single unifying concept. A list of potent twentieth century dualities was provided in chapter 4.

The previous steps provide the clue to identifying the central duality. The master principle of general relativity identifies the comparable role of spatial frames of reference. In step 2 of the map, we also noticed the equivalence of acceleration and matter-sourced gravitational fields. Thus general relativity chose as its duality the previously disparate concepts of space and matter. Note that special relativity had already merged both energy with mass and space with time. In classical mechanics, space was regarded as a given concept: something necessary for matter but not vice versa. Some people had argued that expanses of space could not be said to exist without any matter or energy to provide a measurement scale. However, these dual concepts (space and matter) were so unrelated that neither could be considered to have been thoroughly explored as the dominant partner of the duality. Einstein was able to unify this duality from the perspective of space and he proposed to deploy Riemannian geometry as the basis for the mathematical description of space, matter and gravitation.

16.1.5 Step5: Environment

No theory arises in isolation; yet, some new hypotheses can still be surprising.

General relativity was formulated whilst science was still very uncertain about the nature of the quantum species of theory. However, physics had begun to notice the advantages of field theories in describing elastic materials, fluid motion and electromagnetism. Gauss, Riemann and others had prepared the necessary mathematics and they had even expressed the hope of achieving a geometrical union of space, matter and force. General relativity also clearly gained from the allied understandings made by special relativity in describing space-time and inertial reference frames. Otherwise, like all the great inspirational leaps of imagination, general relativity was a clear break from most contemporaneous thought.

16.1.6 Step 6: Assembly of Picture

Classical mechanics describes a world of definite laws, of cause and effect, of determinism and of certainty. From the above steps of the map, a new picture emerges of the world of general relativity, a picture that retains such classical features. However, this picture also replaces the classical picture of spatial-point particles interacting at a distance. From step 4 we uncover a new foundation for the world described by a space-time metric rather than point particles. Step 2 tells us that gravity and acceleration are to be treated as equivalent frame-dependent descriptions of a common phenomenon, ie the metric describes space, matter in space and geodesic motion. Step 5 tells us to use field equations to describe matter and interactions. Step 3 places constraints on the mathematical representations that must describe everything.

16.1.7 Map Summary

The presented analysis of the steps in the transition from the worldview of classical mechanics (incorporating special relativity) to the worldview of general relativity has been summarized in table 16.1. We have demonstrated the hypothetical value of the map in a historical development of general relativity. Note that the same map has been shown to be useful in initiating the development of quantum mechanics, the flat-space form of eventual reality and general relativity. Consequently these steps appear to provide a classification scheme identifying the individual features that should be worth considering when specifying or developing any new scientific worldview.

Hence let us now turn our attention to the picture that emerges when we apply this same map to the transition from general relativity to a curved-space form of eventual reality.

Table 16.1

Summary of the steps in the map describing the development of general relativity (GR) from classical mechanics (CM).

#	Title	Action	Result (CM→GR)
1	Map.	Copy previous map	
2	Dark Clouds	New additional concepts	Equivalence gravity & acceleration
3	Master Principle	Universalize addition	Invariance of laws of nature
4	Dualities	Concepts resisting merger	Space and matter
5	Environment	Trends, competitors, allies	Field theory
6	Picture	Assemble around elemental	Space geometry
7	Hypothesis	First statement	Equation of space, matter and gravity
8	Presentation	Report on work	Community acceptance

16.2 First Steps beyond General Relativity

It could be argued that general relativity was invented before it was needed. Laboratory-scale classical mechanics was threatened by an impressive array of quantum-type problems before classical mechanics surrendered; whereas, celestial-scale classical mechanics encountered fewer relativistic problems before the celestial classical worldview crumbled when faced by the might of general relativity. If general relativity had not been developed until the mid twentieth century, most probably this document that you are reading would very different. Hence, let us imagine for a moment that we have not completed part 2 of this work. Further, let us imagine that instead of using the map to engineer a hypothetical antecedent of quantum mechanics, we use the same map to generate a hypothetical antecedent of general relativity. It could be argued that we did not set out (in part 2) to develop an antecedent for quantum mechanics. However, we did effectively approach the paradigm-changing process in this way and hence we made an early flat-space approximation. It transpires that from the different perspective of general relativity, we will naturally make different approximations. Of course there is much commonality in these two approaches and hence general relativity's antecedent will look remarkable like a form of the description of eventual reality (on an appropriate storey) that we have already developed. Furthermore, we will expect the analysis to reveal an approximation that directly yields the experimentally tested formalism of general relativity. Considerations of this formalism will be left to the next chapter.

Now, let us exploit the principles of the map to progress beyond the theory of general relativity. We shall start from the worldview of general relativity and we shall follow the first six steps of the map to develop an improved worldview picture. Incidentally, since general relativity retains so many classical precepts, general relativity can be said to be further away from the foundations of science than quantum mechanics. Hence the map may take us on a shorter journey than that journey followed in parts 1 and 2 ie the map's execution may not take us all the way to the foundation of reality.

16.2.1 Step 1: The Map

Without serious update, the map developed in part 1 has been shown to be capable of developing quantum mechanics, the flat-space version of eventual relativity (table 6.1) and general relativity (table 16.1).

Let us therefore adopt this same map again and consider the first six steps from the standpoint of general relativity. For readers who have appreciated the function of the map, these steps will come easily. Thus the explanations and discussions are shorter than those provided in part 1. As usual, greatest attention is paid to the issues of step 2 than to steps 3-6.

16.2.2 Step 2: Dark Clouds

We shall start by considering some of the questionable internal troubles within the theory of general relativity. Now, the grand aim of general relativity – as betrayed by the theory's name – was to create a four-dimensional purely geometrical representation of a mechanistic world of matter, a world that is subjected to geometrically based interactions that mediate changes in motions. The theory has many proven successes; however, this grand aim has been unsuccessful. Matter and interactions have not been fully described geometrically. Furthermore, despite being founded upon the elegant principle of the universality of natural laws, the totality of general relativity has only a limited *a priori* justification and limited applicability. Indeed the lack of full philosophical foundation is revealed in many ways. For example, a large number of elements of the theory are posited: the geometrical principles, the number forms, the dimensionality, the energy tensor, etc.

The major synthesis of general relativity is achieved by the unification of matter and space. The expectation at the centre of this synthesis is that the geometry of space dictates all matter's attributes and in turn, matter imposes space's geometry. Whilst this aim has been demonstrated for the interaction of gravity, when more than one force is active, the aim cannot be achieved within the framework of general relativity without some apparently contrived mathematical device[1].

Indeed were all forces to be somehow equivalently founded in the common geometry of space, a conceptual quandary would be found in the definition of an inertial reference frame. An observer is in an inertial reference frame, if the observer is not subjected to external forces. Unfortunately the phrase 'no external force' cannot mean a zero net force. In general relativity (with a space possessing a single affine connexion), the absence of forces is a subjective (frame dependent) criterion not an objective (frame independent) criterion. An astronaut in non-powered orbit around an isolated planet feels no external forces and therefore the astronaut is in an inertial reference frame despite appearing (from the surface) to be accelerating towards the planet under the command of the force of gravity. In contrast, an observer standing still on the solid surface of the non-spinning rocky planet, experiences no acceleration yet (due to the balance of forces) the observer is not in an inertial frame.

Furthermore the philosophy of general relativity implies that it should be pointless to distinguish matter from the field(s) created by matter. Yet, the fundamental equation of general relativity makes this separation. If there is meaning to the concept of a gravitational field (as quantified by the curvature) in matter-free space, then matter cannot only be a function of the geometry of regular space.

Now, in addition to the internal quibbles just listed, there are two major external inconsistencies that bedevil general relativity. The first problem is the requirement of causal determinism, a requirement that was inherited from classical mechanics. In one form or another, this issue of determinism is prevalent in approaches that equate absolute reality with quantitative *analytical* mathematics. As previously discussed, this inadequacy in the philosophical foundation[2] shows that the prediction of general relativity can only be approximately correct.

The second external problem is shown by the dramatic clash of values that occurs between the principles of the theories of general relativity and of quantum mechanics. These theories represent the two major theoretical innovations of the last century. Logically there can only be one fundamental theory of nature and that fundamental theory can be neither quantum mechanics nor general relativity (both are philosophically unsound theories). This clash of the worldviews of our leading yet incompatible theories is the topic for which we most clearly need a new additional uniting principle. To illustrate this clash, a list of several incompatible features is given in table 16.2.

Table 16.2

List of characteristic differences between the worldviews of quantum mechanics (as derived in part 2) and of general relativity.

Explained Feature	Quantum Mechanics (sourced from eventual reality in part 2)	General Relativity
Deterministic	No	Yes
Structure of space	Particle configuration space with factorable 8 branches & 8 coordinate dimensions	4-dimensional space-time continuum, infinitesimally Lorentzian.
Nature of force	Exchange of virtual particles	Curvature of space-time continuum
Form of mathematics	Based upon independent division-ring symmetries	Tensor algebra
Attributed invariance	Lorentz invariance across inertial frames	Frame invariance of laws of nature
Presumed scope	Laboratory scale matter	Cosmological scale, strong gravity/acceleration

Since the Copenhagen interpretation of quantum mechanics has its own internal disputes, in this discussion we shall let the robust form of quantum mechanics (as sourced in part 2 of this work) provide the tabulated value structure for the theory of quantum mechanics.

Whereas general relativity is notoriously different from the Copenhagen interpretation of quantum mechanics, eventual reality provides quantum mechanics with an underpinning spatial structure. Thus, despite the large number of differences exhibited in table 16.2, a merger of the two views is easily envisaged.

Quantum mechanics can be regarded as the arbitrator of both empiricism and logic. A compatible merging of quantum mechanics and general relativity will provide a path for the full rational justification of general relativity as an approximation on its appropriate storey of reality. Considering how this merger should be achieved highlights the most pressing omission (dark-cloud addition) in the worldview of general relativity.

General relativity has a rationale for several of its features such as general relativity's frame invariance and the curvature of space. Approximations employed in part 2 have led to these features being ignored in eventual reality's quantum mechanics. Obviously it would be useful to import these ideas into quantum mechanics and eventual reality. Alternatively quantum mechanics (sourced from eventual reality) can justify the formal structure of its pan-dimensional space continuum; whereas, general relativity relies on dubious empiricism to justify the form of its space continuum.

Thus the most definite missing ingredient in general relativity is the appropriate description of the spatial continuum. Hence adopting an 8-dimensioned 8-branched configuration space representation (as justified in part 2) will be selected as the appropriate dark-cloud addition of step 2.

16.2.3 Step 3: Master Principle

We are instructed to generate the master principle by further generalizing the additional concept associated with the dark-cloud topic. With the benefit of the conclusions of part 2, we can take this generalization (adoption of eventual reality's space) to its ultimate conclusion. Thus we call for the removal of all empiricisms within the new evolved general relativity by the adoption of eventual reality's *a priori* rationalism. Recall that empiricism enters general relativity through the number form describing the spatial dimensions, the dominance of a metric form, the use of inner products and Newtonian laws embedded in the energy tensor. These contributions to the formulation of general relativity are all to be replaced by justifiable alternatives.

It is already clear from the flat-space form of eventual reality and the discussions of the previous chapter how these remnants of arbitrariness can be expunged from within the framework of general relativity. This replacement will be accomplished in the next chapter.

16.2.4 Step 4: Dualities

A list of powerful dualities was provided in chapter 4. We will select a duality, from this list, for the current analysis. Now, we have identified the key to advancing the paradigm of general relativity as healing the philosophical and conceptual rift between general relativity and quantum mechanics. Quantum mechanics provides only statistical predictions whereas general relativity (like classical mechanics) claims complete certainty in its predictions. Thus, in this treatment, the duality that most interests us is duality 2, which contrasts statistical imprecision and definite exactitude.

This duality has already emerged unexpectedly in other conventional analyses eg black-hole studies. When quantum mechanics is applied locally in the exterior environment above the horizon of a large black hole, one finds that black holes behave as thermodynamics bodies. Black holes are still totally absorptive of radiation but, like thermodynamic blackbodies, black holes are also equally emissive; black holes radiate energy through a mechanism involving the quantum creation of virtual-particle pairs. Some of these virtual particles[3] can pick up sufficient energy from the gravitational field to become real with one particle irretrievably falling into the hole and the other miraculously escaping the gravitational clutches of the hole. This association of the theories of thermodynamics, quantum mechanics and general relativity begs questions about the place and character of statistical order in the greater scheme of things.

Now, general relativity barracks for determinism and exactitude. Thus by calling upon the theory of difficult change, we expect to conjure the union of these clashing concepts, by taking the contrary viewpoint of statistical imprecision. Not surprisingly this is the conclusion that eventual reality has already shown us. Eventual reality is founded upon a chaotic clamour that seethes in the apeiron.

16.2.5 Step 5: Environment

Step 5 deals with what we can learn or import from other current theories and from trends in theoretical developments. There are many proposed half-formulated advanced theories of reality. We have mentioned a few in earlier sections. However, it would be a distraction to describe or even list these alternative viewpoints here; an interested reader can find much detail in the excellent book by Penrose[4]. However, undoubtedly in the field of mature ideas, we must view the part-2 form of quantum mechanics (eventual reality) as the dominant player in influencing the development of a new version of general relativity. The version of quantum mechanics (sourced from eventual-reality) provides the clear *a priori* mathematics needed to fully describe matter. From the environment of ideas, we will conclude that an alliance between general relativity and this new mathematics is clearly needed.

We may view this alliance as the next logical step in the most consistent scientific trend in recorded history. Let us recall how scholarly thinkers have sought explanations for the link between matter and gravity.

The classical Greeks believed that matter was mystically motivated to travel towards its place of equilibrium in a universal stratification characterized by density. For instance, on dropping wood and stone into a pond we observe that the stone sinks yet the wood floats. Furthermore the Greeks believed that it was self-evident that heavier bodies should fall faster than lighter bodies. After all, if we place a heavy object in one hand and a light object in the other hand, then we can feel the difference in the downward forces. In addition, everyday experience informs us that a dropped feather takes longer to touch the ground than a stone which is similarly dropped. However, feathers falling in the air take an irregular time and an irregular path – clearly feathers are strongly influenced by other forces in addition to gravity.

Galileo recognized the logical fallacies of these ancient views that confuse density with weight and ignore the influences of the surrounding media. In the absence of a drift current (ie bulk material flow within the environment), Galileo understood that the surrounding medium provided a frictional resistance to motion and a buoyancy upthrust. He understood that we could reduce the consequences of such influences and we could even imagine bodies in the absence of any surrounding medium. Let us cannibalize his many arguments and present a modern regurgitated summary. Suppose we break a cannon ball into two unequal pieces. Then, we strip the Earth of its atmosphere and we drop these two pieces of ball with different mass from the top of the leaning tower of Pisa. The only relevant force, in their fall to the ground, is that of gravity. Thus any difference in the times taken to fall to the ground can only depend upon the masses of the ball pieces. Imagine that one piece of ball falls faster than the other piece. We repeat the experiment but now we link the ball pieces with a light-weight inextensible rope. The rope can pull but not push upon the ball pieces. The rope will act to slow the faster piece and speed the slower piece so that the two pieces fall together at an intermediate speed. Iteratively we repeat the experiment; each time we shorten the rope until eventually the two pieces are tied closely together as one complete unit with a mass greater than that of either piece individually. Clearly this composite ball cannot fall more quickly than either constituent piece nor can it fall more slowly than either constituent piece. Consequently, as any percentage split of the canon ball is possible, everything that we drop from the top of the tower will fall at the same rate.

Newton was heavily influenced by Galileo's ideas. Newton was able to complete the progress of metamorphosing the quantitative part of natural philosophy into science and he crystallized the precise form for the key conceptual elements that describe the classical mechanistic world. Newton defined mass as the measure of the quantity of substance. He defined momentum as the measure of motion and he defined force as the measure of the rate of change of motion (ie rate of change of momentum). Then, Newton invented the mathematical method of fluxions (calculus) so as to compute gravitational orbits and hence verify a proposed mathematical form for gravity that fitted the experimental planetary data. Hence Newton showed that gravity was universal – the same force that held the Moon in its orbit and held each of the planets in their orbits around the Sun, also controlled the falling of an apple from an apple tree. Now, with mass defined as the measure of the quantity of substance, the occurrence of mass as the source of both inertia and of gravitation was unnoteworthy. Only after other universal forces became understood[5] did the unique role of gravity (as exemplified by the equivalence of inertial and gravitational masses) become seen to be remarkable. Classical mechanics could shed no light on this special role for gravity.

Further progress was not made until Einstein reconsidered this equivalence. Einstein's investigations led him to devise the theory of general relativity. Special relativity had already introduced the concept of space-time (or 4-space). Einstein recognized that introducing the concept of force unnecessarily complicated the computation of trajectories. In a fixed gravitational field, all bodies (with the same initial conditions) follow the same trajectory (ie they are subject to the same accelerations). These trajectories are universal and essentially define the observable structure of space-time. This structure defines the mathematical distortion (of 4-space from Euclidean flatness) that is identified with the gravitational field. However, the theory of general relativity is incomplete. General relativity incorporates gravity but not the other forces. Similarly general relativity incorporates some features of matter but not all features (such as their quantum numbers). Since matter is the source of gravity, the incompleteness of the description of matter is a serious deficiency.

However, this inadequacy of general relativity is neatly matched by that of quantum mechanics which claims to adequately describe matter but not space. Thus it is recognized that the next stage in this progression is to conflate the theories of quantum mechanics and general relativity. Many have tried, yet this conflation has not been accomplished probably because both conventional theories are in need of substantial reformation. Part 2 of this work has shown how the reformation of quantum mechanics is achieved. Hence this earlier work can inform the reformation of general relativity. Then, we shall be able to locate both theories on appropriate storeys of reality as useful descriptions and approximations.

16.2.6 Step 6: Picture

As we did previously, we may assemble the results of these early steps of the map to form a new perspective on reality. This perspective is summarized in table 16.3. This final picture revealed by the current map has been strongly influenced by the precepts of eventual reality (as deduced by a previous journey undertaken using the same map). Nevertheless it is worth describing the assembly of pieces to reveal the sought-after prize – a vision of a form of eventual reality coloured by general relativity.

The steps in the map suggest that there is an ultimate statistical foundation (step 4) lying beneath the recognizable space-based reality of general relativity. The steps imply that we should construct a description of this space-based reality by re-dimensioning space (step 2); we formally attribute to space an 8x8 structure of directions and branches with additional labelling for each particle. Step 5 suggests that we adopt (from the earlier eventual reality) new expressions to describe matter (eg replace the energy tensor by statistical information). These descriptions of matter should be based upon self-justifiable number forms (step 3). These new interpretations will have their origin in the reality underlying space.

This picture will be given a mathematical formalism in the next chapter.

Table 16.3

Table summarizing the first six steps of the map that shows the transitions from quantum mechanics (QM) to eventual reality (ER), classical mechanics (CM) to general relativity (GR) and general relativity to a new view of eventual reality.

#	Title	QM→ER	CM→GR	GR→ER
1	Map.	Update		
2	Dark Clouds	Chaos, complex numbers	Equivalence gravity & acceleration	8x8 Hilbert space
3	Master Principle	Unspecialized numbers	Invariance of laws of nature	Number forms from eventual reality
4	Dualities	Quality and quantity Mind and matter	Space and matter	Statistical imprecision and exactitude
5	Environment	Disorder, Fermat's principle	Field theory	Quantum matter
6	Picture	Qualitative foundation	Metrical base	Statistical foundation

Notes

[1] The first such adjunct device was the fifth dimension introduced by Kaluza to permit both electromagnetism and gravity.

[2] Determinism is philosophically unsound. Thus, as discussed in section 8.2, we regard deterministic theories as approximations.

[3] Virtual particles have no long-time (average) existence. They do not satisfy the classical relations between energy, momentum and velocity uncovered in section 11.1; they necessarily have an energy debt.

[4] Penrose (2004) provides a clear introduction to the mathematical principles that are being currently being used to cautiously probe reality.

[5] In the 1860s, James Clerk Maxwell showed that electrical and magnetic-induction fields were different aspects of a single electromagnetic field and that light was electromagnetic radiation.

Part 3

Chapter 17 : Relativistic Formalism of Eventual Reality

Step 7 of the map requests that, in the first instance of the development, we make suitable approximations and be willing to retain some familiar concepts. In part 2 we retained flat space; in part 3 we are allowing a curved space but we shall retain the classical limit through certain simplifications.

Matter-related distortions of space will not generally be the same on the different branches. However, distortions are well-matched across the branches in the classical limit. This common representation of space is one of the reasons why the classical limit is particularly easy to handle. On the classical level of reality, a classical part of a system is made up from approximately the same large number[1] (eg $> \exp(10)$) of fundamental particle fields on each branch. Furthermore, on the inspection scale relevant to (large-scale) classical resolution these particle fields are closely located and share a common variability. Statistical fluctuations of particle numbers will not influence the impression of an unchanging definite classical reality. For instance, the effectively weak coupling to stray random fields (eg due to background illumination) has no noticeable effect upon the branch occupancy and location of particles.

We will now investigate how we may reformulate the matter-storey of eventual reality for three different-style spaces. We plan to drop the flat-space approximation and we shall consider variations in the metric of space as well as particle-field fluctuations. Thus we need to render compatible and to integrate the two systems of mathematics that we have been using.

17.1 Lorentzian Space

We will start by recapitulating how we derived quantum mechanics from a variational equation. However, this time we shall use tensor notation. In its simplest form we wrote the entropy parameter on one branch in Lorentzian space as

$$\underline{\underline{S}} = \int d\tilde{q}\, \underline{\underline{\Lambda}} \qquad (17.1)$$

where $\underline{\underline{\Lambda}}$ is second order in single-particle fields, $\underline{\underline{\phi}}$. We shall ignore branch indices and add the contributions from the different directions, ie

$$S = \sum_{\lambda} S_{\lambda} \qquad \text{and} \qquad \Lambda = \sum_{\lambda} \Lambda_{\lambda} \cdot \tag{17.2}$$

Then using tensor notation we may write

$$S = \int d\tilde{q}\,\Lambda \cdot \tag{17.3}$$

For a single isolated free (ie non-interacting) particle we rewrite equation (10.81) in invariant form[2] as

$$\Lambda = \tfrac{1}{2}\sum_{jk} c^{j}\left(\phi^{k}\phi_{k}\right)_{,j} \cdot \tag{17.4}$$

The equation (17.3) is not yet fully tensorial – the integral must also be made appropriately scalable. Now, the metric equation for the infinitesimal invariant interval is

$$ds^{2} = \sum_{ij} dq_{i} g_{ij} dq_{j} \cdot \tag{17.5}$$

In a coordinate representation in which the matrix $g_{\mu\nu}$ is diagonal, the diagonal elements are the eigenvalues. Each infinitesimal length, dq_i, will scale with the inverse of the square roots of its associated eigenvalue. Thus the invariant volume element under scaling is $d\tilde{q}\sqrt{\|g_{\mu\nu}\|}$, where $\|g_{\mu\nu}\|$ is the determinant (product of eigenvalues) of the matrix with elements $g_{\mu\nu}$. Hence, we can express the quantum-matter contribution to the entropy in an invariant manner (under coordinate-frame transformations) by adopting the tensor-density term

$$S[g_{ij}, \phi_k] = \int d\tilde{q}\sqrt{\|g_{\mu\nu}\|}\,\Lambda \cdot \tag{17.6}$$

Both the metric of space and the particle fields of matter may be independently varied to generate a variational equation of the form

$$\delta S = \int d\tilde{q}\sqrt{\|g_{\mu\nu}\|}\left(-\sum_{ij} T_{ij}\delta g^{ij} + \sum_{i} U^{i}\delta\phi_{i}\right) \cdot \tag{17.7}$$

The application of Fermat's principle requires that the entropy parameter be stationary for any changes to the network. Then considering particle field variations, we set $U^{i} = 0$ and obtain the unique form for Λ. This is the form that we depicted in equation (17.4) and is confirmed by the result

$$U^{i} = -\sum_{jkl} c_{k} g^{kj}{}_{;j} g^{li}\phi_{l} = 0 \cdot \tag{17.8}$$

Now, in the evaluation of the determinant of the metric, the cofactor of an element, g_{ij}, can be expressed as $\|g_{\mu\nu}\|g^{ij}$. The cofactor is independent of g_{ij} and hence

$$\delta\|g_{\mu\nu}\| = \|g_{\mu\nu}\|\sum_{ij} g^{ij}\delta g_{ij} = -\|g_{\mu\nu}\|\sum_{ij} g_{ij}\delta g^{ij}\,. \qquad (17.9)$$

Thus

$$\delta\sqrt{\|g_{\mu\nu}\|} \equiv \frac{\delta\|g_{\mu\nu}\|}{2\sqrt{\|g_{\mu\nu}\|}} = -\tfrac{1}{2}\sqrt{\|g_{\mu\nu}\|}\sum_{ij} g_{ij}\delta g^{ij}\,. \qquad (17.10)$$

The stationariness of the metric variation[3] in equation (17.7) requires $T_{ij} = 0$ ie

$$T_{ij} = \tfrac{1}{2}\Lambda g_{ij} + \tfrac{1}{2}\phi_i\phi_j\sum_{kl} c_k g^{kl}{}_{;l} = 0\,. \qquad (17.11)$$

This result provides us with the condition that Λ vanishes throughout all space - a result that was justified less rigorously in part 2. Hence

$$\Lambda = \tfrac{1}{2}\sum_{jk} c^j\left(\phi^k\phi_k\right)_{;j} = 0\,. \qquad (17.12)$$

Thereafter the algebra of the fixed coefficients, c^j, is determined by universal symmetries[4]. This equation (17.12) is the non-interacting-particle equation of motion.

17.2 Riemann Space

Next, let us allow our space to be curved. We know that all global equations for infinitesimal quantities must factor into equations using variables displaying Q_8 and SU(2) symmetries. Hence the manifold of space is factorable into one space of four dimensions and one space of two dimensions. In the four-dimensional space we can have a parameterized curvature eg a curvature at a point that varies in time. However, such a parameterization is not possible for the two-dimensional space. The geometry of the two-dimensional space is frozen (unparameterized) and apeironic translational invariance ensures that the two-dimensional spatial metric describes a flat 2-D space. Thus we shall only consider a four-dimensional *reduced* space.

Next, we need to add a further term to represent the information pertaining to the curvature of the spatial framework.

$$S = \int d^4X \sqrt{\|g_{\mu\nu}\|}\left(\Lambda(\{\phi_i\},\{\phi_{j;k}\}) + C/4\pi\alpha^2\right) \qquad (17.13)$$

The metric's contributions that are associated with quantum matter have already been included within Λ which has been rescaled due to the change in unit designation (from dimensionless to dimensioned variables). C represents an additional term that is derived solely from the distortion from flatness of space. Thus there is a contribution from the curvature of space (leading to a gravitational field) that is separate from the matter-particle probability field. The additional term must be representable independently of any specific coordinate system ie be a zero-rank tensor. This term must vanish for flat space. This term can depend only upon the metric. Hence C should be the simplest analytic invariant zero-rank tensor with a covariant weight of zero. These requirements for C are met by the Ricci scalar of curvature[5]. The numerical-real factor, α, calibrates the quantity of information within our unit system ie relates the event size to the system of measurement units. We may deduce the magnitude of α from the coordinate power (table 16.3) of the Ricci scalar. R has a coordinate power of –2. For S to have coordinate power of zero, α must have a coordinate power of one. The only applicable scale is that provided by the size attributed to an event. Thus we should not be surprised that α is proportional to the event's size.

$$S = \int d^4X \sqrt{\|g_{\mu\nu}\|}\left(\Lambda(\{\phi_i\},\{\phi_{j;k}\}) + R/4\pi\alpha^2\right) \qquad (17.14)$$

From our definition of $R^i{}_{jkl}$ (equation 15.54) we have

$$\delta R_{ij} = \sum_k \delta R^k{}_{ikj} = \sum_k \left(\delta\Gamma^k{}_{ij;k} - \delta\Gamma^k{}_{ik;j}\right). \qquad (17.15)$$

Multiplying equation (17.15) by g^{ij}, integrating the covariant-derivative terms by parts and recognizing that the covariant derivative of g_{ij} is zero, we obtain under the integral

$$\sum_{ij} g^{ij}\delta R_{ij} = 0. \qquad (17.16)$$

More generally under the integral,

$$\delta\left(R\sqrt{\|g_{\mu\nu}\|}\right) = \sum_{ij}\delta\left(R_{ij}g^{ij}\sqrt{\|g_{\mu\nu}\|}\right) = \sum_{ij} R_{ij}\delta\left(g^{ij}\sqrt{\|g_{\mu\nu}\|}\right). \qquad (17.17)$$

Now,

$$\delta\left(g^{ij}\sqrt{\|g_{\mu\nu}\|}\right) = \sqrt{\|g_{\mu\nu}\|}\left(\delta g^{ij} - \tfrac{1}{2} g^{ij}\sum_{mn} g_{mn}\delta g^{mn}\right). \qquad (17.18)$$

Hence, we obtain the expression for the variation in the curvature term

$$\sum_{ij} R_{ij}\delta\left(g^{ij}\sqrt{\|g_{\mu\nu}\|}\right) = \sqrt{\|g_{\mu\nu}\|}\left(\sum_{ij}\delta g^{ij}R_{ij} - \tfrac{1}{2}R\sum_{mn} g_{mn}\delta g^{mn}\right) \quad (17.19)$$

$$= \sqrt{\|g_{\mu\nu}\|}\sum_{ij}\delta g^{ij}\left(R_{ij} - \tfrac{1}{2}Rg_{ij}\right). \quad (17.20)$$

Consequently the functional derivative of the entropy factor with respect to variations of the metric is evaluated as

$$\frac{\delta S[g^{ij},\phi]}{\delta g^{ij}} = \sqrt{\|g_{\mu\nu}\|}\left(-T_{ij} + (R_{ij} - \tfrac{1}{2}g_{ij}R)/4\pi\alpha^2\right). \quad (17.21)$$

The entropy factor is required to be stationary; the functional differential must vanish. Hence, we arrive at the dynamic tensor equation relating space and matter

$$-4\pi\alpha^2 T_{ij} + (R_{ij} - \tfrac{1}{2}g_{ij}R) = 0. \quad (17.22)$$

This equation is the principle tensor equation of general relativity:

$$G_{ij} = R_{ij} - \tfrac{1}{2}Rg_{ij} = 4\pi\alpha^2 T_{ij}. \quad (17.23)$$

This equation (17.23) has an alternative form which we can uncover by performing the operation $\sum_{ij} g^{ij}$. Then

$$-R = 4\pi\alpha^2 T \qquad \text{where} \qquad T = \sum_{ij} g^{ij}T_{ij} \quad (17.24)$$

and hence the alterative form becomes

$$R_{ij} = 4\pi\alpha^2(T_{ij} - \tfrac{1}{2}Tg_{ij}). \quad (17.25)$$

Next, we shall confirm that the strength of the gravitational curvature terms is determined by the (Planck) size scale attributed to an event. Gravity is a weak force (today) because the scale of an event is so small in our system of measurement units.

17.2.1 Geodesics

Geodesic lines are the invariant stationary paths in curved space that infinitesimal particles (eg elementary quanta) would follow in the absence of non-gravitational forces. On the infinitesimal scale, a variation in the path leaves the interval unchanged

$$\delta_{w} \int_{P_0}^{P_1} ds = \int_{0}^{1} d\lambda \delta w = 0 \cdot \qquad (17.26)$$

The variation is taken over a path, w, between fixed points, P_0 and $P_1 \cdot \lambda$ parameterizes the fractional position the along the path.

The geodesic equation enables us to examine the trajectory that a particle would take in a gravitational field. Hence, in standard fashion we can show that in first order the tensor equation reduces to Newton's gravitational formula.

From the metric equation we have

$$w^2 d\lambda^2 = \sum_{ij} dX_i g_{ij} dX_j \qquad (17.27)$$

$$2w\delta w = \sum_{ij}\left(\sum_{k}\left(\frac{dX_i}{d\lambda}\frac{dX_j}{d\lambda}\frac{\partial g_{ij}}{\partial q_k}\delta X_k\right) + 2g_{ij}\frac{dX_i}{d\lambda}\delta\frac{dX_j}{d\lambda}\right) \cdot \qquad (17.28)$$

Integrating by parts

$$\int_0^1 d\lambda \delta w = \int_0^1 d\lambda \sum_{ijk} \frac{\delta X_k}{2w}\left(\frac{dX_i}{d\lambda}\frac{dX_j}{d\lambda}\frac{\partial g_{ij}}{\partial X_k} - 2\delta_j^k \frac{d}{d\lambda} g_{ij}\frac{dX_i}{d\lambda}\right) \qquad (17.29)$$

$$= \int_0^1 d\lambda \sum_{ijk} \frac{w\delta X_k}{2}\left(\frac{dX_i}{ds}\frac{dX_j}{ds}\frac{\partial g_{ij}}{\partial X_k} - 2\delta_j^k \frac{d}{ds} g_{ij}\frac{dX_i}{ds} - 2\delta_j^k g_{ij}\frac{d^2X_j}{ds^2}\right)$$

$$= \int_0^1 d\lambda \sum_{ijk} \tfrac{1}{2} w\,\delta q_k \left(\frac{dX_i}{ds}\frac{dX_j}{ds}\frac{\partial g_{ij}}{\partial X_k} - 2\frac{\partial g_{ik}}{\partial X_j}\frac{dX_j}{ds}\frac{dX_i}{ds} - 2\delta_j^k g_{ij}\frac{d^2X_j}{ds^2}\right) \qquad (17.30)$$

$$= 0\,. \qquad (17.31)$$

We express this result in a symmetric form as

$$\begin{aligned} = \int_0^1 d\lambda \sum_{ijk} \tfrac{1}{2} w\delta q_k \Bigg(& \frac{dX_i}{ds}\frac{dX_j}{ds}\frac{\partial g_{ij}}{\partial X_k} - \frac{\partial g_{ik}}{\partial X_j}\frac{dX_j}{ds}\frac{dX_i}{ds} \\ & - \frac{\partial g_{jk}}{\partial X_i}\frac{dX_j}{ds}\frac{dX_i}{ds} - 2\delta_j^k g_{ij}\frac{d^2X_j}{ds^2}\Bigg) \cdot \end{aligned} \qquad (17.32)$$

If this equation is to hold quite generally for any metric, then we should assume that the following equation holds individually for every i and every j.

$$g_{ij}\frac{d^2X_j}{ds^2} = \tfrac{1}{2}\sum_k\left(\frac{dX_i}{ds}\frac{dX_j}{ds}\frac{\partial g_{ij}}{\partial X_k} - \frac{\partial g_{ik}}{\partial X_j}\frac{\partial X_i}{ds}\frac{dX_j}{ds} - \frac{\partial g_{jk}}{\partial X_i}\frac{\partial X_i}{ds}\frac{dX_j}{ds}\right) \qquad (17.33)$$

Next, we multiply though by g^{im} to obtain

$$\delta_j^m\frac{d^2X_j}{ds^2} = \tfrac{1}{2}\sum_{ik}g^{im}\left(\frac{dX_i}{ds}\frac{dX_j}{ds}\frac{\partial g_{ij}}{\partial X_k} - \frac{\partial g_{ik}}{\partial X_j}\frac{\partial X_i}{ds}\frac{dX_j}{ds} - \frac{\partial g_{jk}}{\partial X_i}\frac{\partial X_i}{ds}\frac{dX_j}{ds}\right). \qquad (17.34)$$

Hence, we obtain the geometric differential equation for a geodesic

$$\frac{d^2X_j}{ds^2} = -\Gamma^j{}_{ik}\frac{dX_i}{ds}\frac{dX_k}{ds}. \qquad (17.35)$$

In the slow weak-field classical limit, our test particle experiences small velocities and small accelerations. With little changing, the factors dX_i/ds are all small except for the temporal term, which is approximately unity. Then, in the first approximation we write

$$\frac{d^2X_j}{dX_0} \approx -\Gamma^j{}_{00} \qquad j = 1\ldots3. \qquad (17.36)$$

If g_{ij} varies with other dimensions but g_{ij} is constant in time, then $\Gamma^j{}_{00}$ approximates to

$$\Gamma^j{}_{00} \approx -\tfrac{1}{2}\sum_k g^{jk}\frac{\partial g_{00}}{\partial X_k}. \qquad (17.37)$$

For the spatial directions we approximate

$$g^{jk} = -\delta^{jk} \qquad j = 1\ldots3. \qquad (17.38)$$

Hence, we note that to a first approximation $\tfrac{1}{2}c^2 g_{00}$ plays the role of the gravitational potential of a distant (weakly attractive) immobile source like the Sun

$$\frac{d^2q_j}{dX_0^2} \approx -\tfrac{1}{2}\frac{\partial g_{00}}{\partial X_j} \text{ for } j = 1\ldots3. \qquad (17.39)$$

Now, we turn to terms in the main tensor equation. With little changing in time

$$R_{00} = \Gamma^{k}{}_{00;k} - \Gamma^{k}{}_{0k;0} \approx \Gamma^{k}{}_{00;k} \,. \qquad (17.40)$$

Thus we find that the Ricci curvature tensor approximates to del squared (rank-zero second derivative) of the gravitational potential

$$R_{00} \approx -\tfrac{1}{2}\sum_{k} g^{kk}\frac{\partial^2 g_{00}}{\partial X_k^2}\,. \qquad (17.41)$$

Now, in the classical slow-moving limit, we may replace all differentials with their corresponding Fourier conjugates. The time-differential contribution delivers an energy term and the spatial derivative contributions deliver momentum terms. Hence

$$T_{ij} \sim \frac{\mathrm{E}_{ij}}{\hbar c}\,. \qquad (17.42)$$

Thus in a dimensional unit system, T_{ij} is proportional to an energy density, E_{ij}. Now, for a static field we ignore momentum terms and keep only the E_{00} term.

Observe that the Einstein tensor equation (17.25) approximates to

$$R_{00} \approx 2\pi\alpha^2 T_{00} \qquad (17.43)$$

$$\nabla^2(-\tfrac{1}{2}g_{00}) = R_{00} = 2\pi\alpha^2\frac{\mathrm{E}_{00}}{\hbar c}\,. \qquad (17.44)$$

We shall take E_{00} to be the energy density, $mc^2\delta(X)$, of a point mass, m. Note that away from the point mass the field decays inversely with distance ie using radial coordinates

$$\nabla^2(1/r) = 0 \qquad \text{for} \qquad r \neq 0\,. \qquad (17.45)$$

Hence the right hand side of this equation describes the source of the gravitational field. We use Green's theorem to take a volume integral centred on the point mass

$$\int_V dV\,\nabla^2(1/r) = \int_S dS.\nabla(1/r) = 4\pi r^2(-1/r^2) = -4\pi \,. \qquad (17.46)$$

Consequently equation (17.44) has solution for the gravitational field

$$-\tfrac{1}{2}c^2 g_{00} = -\frac{\alpha^2 c^3}{2\hbar}\frac{m}{r} = -G\frac{m}{r} \qquad (17.47)$$

where we have set α to equal the square root of two multiplied by the Planck-distance scale.

17.3 Non-Riemann Space

We have seen (in part 2 of this document) that our system of measurement units is dependent upon the age of the universe. Furthermore recall that mass is identified as a measure of local information content and recall that the dimensioned quantum mass scale is inherently time dependent. Thus the dimensioned spatial coordinates[6] are interdependent and consequently we do not have a Riemann space. Therefore we need to specify an invariant 1-form to attribute the variation in scale across space and time eg for a single particle we write

$$d\underline{\underline{\chi}} = \frac{H}{c}\left(dX_0\hat{i}\hat{\gamma}_0\hat{e}_F + dX_1\hat{\gamma}_1\hat{e}_F + dX_2\hat{\gamma}_2\hat{e}_F + dX_3\hat{\gamma}_3\hat{e}_F\right). \qquad (17.48)$$

The scale factor is a rank-zero tensor

$$\Theta = \frac{H}{c}. \qquad (17.49)$$

We will generate higher-ranked tensors by covariant differentiation

$$\sigma_i = \frac{\Theta_{;i}}{\Theta} \text{ and } \sigma_{ij} = \sigma_{i;j}. \qquad (17.50)$$

Nowadays the age of the universe is large compared to the unit of atomic time; consequently Θ is only very slowly varying in time and, for many purposes, Θ can almost be regarded as a constant. The scale factor describes the change from the dimensionless q_i to the dimensioned variable X_i

$$dq_i \to \Theta(\widetilde{X})dX_i \qquad (17.51)$$

and comes with an associated renormalization of the metric

$$g_{ij} \to \Theta^{-2}\breve{g}_{ij} \quad \text{and} \quad g^{ij} \to \Theta^2\breve{g}^{ij} \qquad (17.52)$$

where the $\breve{g}_{ij}$ is a scale-free metric ie

$$\breve{g}_{ij;k} = 0. \qquad (17.53)$$

The expanding scale is common to both curved space and to its mapped tangent space. Thus in flat tangent space the metric derivatives differ from zero as

$$g_{ij,k} = -2\sigma_k g_{ij} \quad \text{and} \quad g^{ij}{}_{,k} = 2\sigma_k g^{ij}. \qquad (17.54)$$

In curved space we have the corresponding results for the covariant derivatives

$$g_{ij;k} = -2\sigma_k g_{ij} \qquad \text{and} \qquad g^{ij}{}_{;k} = 2\sigma_k g^{ij}. \tag{17.55}$$

Explicitly using the affinity to evaluate the covariant derivative of the metric yields

$$g_{ik;l} = -2\sigma_l g_{ik} = g_{ik,l} - \sum_m \left(g_{mk}\Gamma^m{}_{il} + g_{im}\Gamma^m{}_{kl}\right). \tag{17.56}$$

We reorganize this equation as

$$-g_{ik,l} - 2\sigma_l g_{ik} = -\sum_m \left(g_{mk}\Gamma^m{}_{il} + g_{im}\Gamma^m{}_{kl}\right). \tag{17.57}$$

The cyclic permutation of indices gives

$$g_{kl,i} + 2\sigma_i g_{kl} = \sum_m \left(g_{ml}\Gamma^m{}_{ki} + g_{km}\Gamma^m{}_{li}\right) \tag{17.58}$$

and

$$g_{li,k} + 2\sigma_k g_{li} = \sum_m \left(g_{mi}\Gamma^m{}_{lk} + g_{lm}\Gamma^m{}_{ik}\right). \tag{17.59}$$

Adding these three equations and using the symmetries of the metric and affinity gives

$$(g_{kl,i} + g_{li,k} - g_{ik,l}) + 2(\sigma_i g_{kl} + \sigma_k g_{li} - \sigma_l g_{ik}) = 2\sum_m g_{ml}\Gamma^m{}_{ik}. \tag{17.60}$$

Next, we multiply through this equation by g^{jl} and sum over index l. Summations commute and we get an explicit expression for the affinity

$$\Gamma^j{}_{ik} = ½\sum_l g^{jl}(g_{kl,i} + g_{li,k} - g_{ik,l} - 2\sigma_l g_{ik}) + (\sigma_i\delta^j_k + \sigma_k\delta^j_i). \tag{17.61}$$

The evaluated affinity is just the Christoffel affinity

$$\begin{Bmatrix} j \\ ik \end{Bmatrix} = ½\sum_l g^{jl}(g_{kl,i} + g_{li,k} - g_{ik,l}) \tag{17.62}$$

plus a small correction ie

$$\mathrm{K}^j{}_{ik} = \left[\sigma_i\delta^j_k + \sigma_k\delta^j_i - g_{ik}\sigma^j\right]. \tag{17.63}$$

$$\Gamma^j{}_{ik} = \begin{Bmatrix} j \\ ik \end{Bmatrix} + \mathrm{K}^j{}_{ik} \tag{17.64}$$

Thus we append a correction to the Einstein-Ricci curvature tensor of

$$R_{ij} = \sum_k (\mathrm{K}^k{}_{ij;k} - \mathrm{K}^k{}_{ik;j}). \tag{17.65}$$

Evaluating this correction to the curvature due to the expansion of the universe

$$R_{ij} = \sum_k \left(\sigma_{ik}\delta^k_j + \sigma_{kj}\delta^k_i + 2g_{ij}\sigma_k\sigma^k - g_{ij}\sigma^k{}_k\right) - 4\sigma_{ij} \qquad (17.66)$$

or

$$R_{ij} = -2\sigma_{ij} + g_{ij}\sum_k \left(2\sigma_k\sigma^k - \sigma^k{}_k\right). \qquad (17.67)$$

This provides an extra contribution to the entropy parameter of

$$S = a\int d^4X\sqrt{\|g_{\mu\nu}\|}\sum_{ijk} g^{ij}R_{ij}\,. \qquad (17.68)$$

Thus

$$S = a\int d^4X\sqrt{\|g_{\mu\nu}\|}\left(4\sum_k (2\sigma_k\sigma^k - \sigma^k{}_k) - 2\sum_{ij} g^{ij}\sigma_{ij}\right) \qquad (17.69)$$

which (with $a = 1/4\pi\alpha^2$) simplifies to

$$S = -6a\int d^4X\sqrt{\|g_{\mu\nu}\|}\sum_{ij} g^{ij}\sigma_{ij}\,. \qquad (17.70)$$

It is useful to define the scalar invariant

$$\Lambda = 3\sum_{ij}\sigma_{ij}g^{ij} \qquad (17.71)$$

for which

$$\delta\Lambda = 0\,. \qquad (17.72)$$

Thus we obtain an additional contribution to a variation in the metric of

$$\delta S = a\int d^4X\sqrt{\|g_{\mu\nu}\|}\sum_{ij}\delta g^{ij}g_{ij}\Lambda \qquad (17.73)$$

and the dynamic equation for matter becomes

$$-4\pi\alpha^2 T_{ij} + (R_{ij} - \tfrac{1}{2}g_{ij}R + \Lambda g_{ij}) = 0\,. \qquad (17.74)$$

Thus the only change from equation (17.22) is the required addition of a small cosmological-constant term for which we have an explicit expression. Evaluating equation (17.71) in the microwave-background frame of the universe

$$\Lambda = \frac{3}{A_t^2} \qquad (17.75)$$

where A_t is the age of the universe. In distance units this parameter has a value $\sim 2\times10^{-52}\text{m}^{-2}$; in time units this parameter has a value $\sim 2\times10^{-35}\,\text{s}^{-2}$ - a value that is decreasing ever so slowly.

Note that the cosmological constant is approximately known from cosmological studies. The value given in equation (17.75) is of an acceptable order of magnitude.

17.4 Cosmology

In this work we are endeavouring to keep our analyses simple. Detailed equations describing quantum-matter, curved-space, elementary particles, etc have been left to later work. Likewise we will only deliver an introductory cosmology here.

Within the standard accepted scientific paradigms prevenient to eventual reality, cosmology was founded upon a dualism. Thus conventional theories of cosmology implicitly place matter and energy as one side of the dualism and they place space as the other side of this dualism. Such traditional cosmologies have failed to answer questions about the origin of the universe, the nature of reality and reasons for any existence. These cosmologies are necessarily limited because they adhere to a quantitative materialistic viewpoint and by default they describe matter within the non-fundamental compacted and reduced hybrid space of our direct experience.

In contrast, the new paradigm provided by eventual reality casts all of *quantitative* reality in terms only of information; eventual reality acknowledges an infinite underlying coordinate space; eventual reality can provide answers to the fundamental questions of reality. Yet, other materialist questions about hybrid space need answering. Without much of an analysis, there are several important conclusions that we may draw.

17.4.1 Space Flatness and Matter Scales

Firstly analytical continuity appears in the full eight dimensions of coordinate space ie both position and specific momentum. The quaternion-vector variables are reciprocally transformed to produce the dimensions of hybrid space. Equivalences of variables in these spaces ensure that a smooth form is instilled in hybrid space. This smooth form enables us to treat positional space as flat on the ultra-small scale. However, the same result applies to momentum space. Thus because of the Fourier relationships between coordinate and hybrid spaces, we deduce that positional space must also be flat on the ultra-large scale. Hence the material content of space does not set the gross curvature of space; however, the organization of matter will affect how matter moves in apparent (reduced) 4-space.

Coordinate space and hybrid space are described by both different algebrae and different systems of units. The unit change accommodates the scale dependent change in descriptions that are built into the Fermat (macro-averaging) analysis.

We observe that the spatial extent of the universe is quantified by a large number, $N_U = cA_t / x_{Planck}$, where x_{Planck} is the unit Planck length. This number represents the count of pointer links (in the stationary sub-networks on each branch) that traverse each dimension. The aethereal stubble grows from each of these events. The transformation to hybrid-space cumulates this stubble growth over spatial regions to assemble the definite but fickle form of the quantum matter. Much larger-scale averaging again produces the certainty of a classical reality.

We shall assume that this aethereal stubble satisfies a multinomial distribution. Then we perform a summation over all the aethereal-stubble values along a line of events. The statistical standard deviation is a low-order number multiplied by factor, $\sqrt{N_U}$. If we substantially exceed the measure of uncertainty provided by the standard deviation, reality will appear definite. Thus on dropping below this distance scale, $\sim x_{Planck}\sqrt{N_U}$, complete certainty must be lost and mathematical probabilities become the order of the day. Today this estimated onset of necessary quantum behaviour (ie failure of the classical description) is at approximately 0.05 mm (0.002"). Hence classical certainty with its strict relationships between variables (eg momentum equals mass times velocity) is a result of large-scale averaging and classical certainty is limited to the large scale. When there is an insufficient data volume available (to provide statistical certainty) many attributions of the aether are described as quantum-virtual. This quantum realm is short on information that can be represented as classically certain. This lack of classical knowledge is genuine, it does not stem from ignorance (as in a hidden variable theory) but it arises from the quantum realm possessing a statistical base that is too-small to display fully classical features. Thus quantum systems can be larger than 0.05 mm but classical systems cannot be much smaller than 0.05 mm.

Matter emerges from a transformation of the quaternion-vector directions. Let us consider this quaternion-vector space and the lower size limit for quantum descriptions. The quantum realm interprets the statistics as information. Thus the quantum realm does not extend down to the non-analyticity of the Planck scale. We may only continue to use quantum mechanics whilst the statistics are reliable. We noted that full classicality requires a 3-volume of at least $x^3_{Planck}N_U^{3/2}$. We see classical information when the summation exceeds the standard deviation, which is the signal noise level, the screech of the virtual dance of matter.

Consequently a 3-volume, $x_{Planck}\sqrt{N_U}$ by x_{Planck} by x_{Planck}, would correspond to the smallest region of 3-space that could retain any classical definiteness. The shape of this 3-volume is unimportant; thus we need a volume of more than approximately $\sqrt{N_U}x^3_{Planck}$ to preserve statistical probabilities. Therefore there is a distance scale, $x_{Planck}N_U^{1/6}$, below which even quantum analyses can start to become inapplicable. We need a larger volume to retain useful classically extractable information. Corresponding to this smallest acceptable quantum distance (2.3×10^{-25} m), we have a smallest acceptable quantum conjugate velocity, $x_{Planck}N_U^{1/6}/(cA_t)$ ie 1.8×10^{-51} of light speed. However, there are four branches of space in each time-sense set. In a quantum analysis we must always match the volumes across each branch. Nevertheless we are not so contained in a sub-quantum analysis. Therefore with 12 quaternion-vector dimensions across these branches, the fundamental information threshold is reached at an estimated distance of $x_{Planck}N_U^{1/24}$ (ie 5.6×10^{-33} m).

17.4.2 Matter Homogeneity and the Discreteness of Space

Gravity has ensured that matter does not appear (to the unassisted eye) to be homogeneously distributed around us in 3-space. Nevertheless on scales larger than galactic-cluster aggregations, net gravitational forces become less contingent upon specific arrangements of matter and we see a fairly uniform distribution of matter. Furthermore the macroscopic noise level attributable to all forms of energy is determined by the era and this noise level is not determined by the direction in which we gaze. Thus the apparent large-scale thermodynamic equilibrium across the universe does not require or imply any thermal contact (eg through the time-reversed branches) between vastly separated regions of 3-space. The global cosmological features of the universe are not determined by the interplay of matter and space; we must look to the underlying aethereal field to explain the properties and evolution of the universe.

Now, eventual reality has shown that hybrid space is created by a collective transformation that is applied to selected finite parts of the infinite universe. Evidentially there is a practical limit (dictated by the age of the universe) to how large this part of the universe can be made. This finite limit should be no more surprising that that there should be a maximum practical speed (ie the speed of light). Let us review the naturalness of this result. The speed of light is introduced as a constant that relates the units of distance and time ie enables us to compare the quaternion-scalar and quaternion-vector sizes. This constant needs to be finite or we will lose dimensions and thereby subvert the required symmetry. The special-relativity transformations follow directly from the metric (eg see equation (15.34)) turning this introduced constant into a speed limit for the 3-velocity. The analytical factorability of coordinate space means that we must also have a limit on the maximum 3-distance for each age. Furthermore we predict this limit of spatial extent to be the product of the age of the universe and the speed of light. Hence the finiteness of revealed space is simply a manifestation of the necessity for symmetry imposed by coherence.

The (large-scale) equivalence of coordinate and hybrid space requires that these spaces are similarly maximally limited in their quaternion-vector subspaces. Hence we deduce a relation between their minimum scales.

Table 17.1

The dimensional 3-vector discrimination limits for distance and velocity coordinates.

	Minimum discrimination	Maximum discrimination
Distance	$2x_p$	$2cA_t$
Velocity	x_p / A_t	c

However, there is an asymmetry between upward-expanding position and downward-contracting velocity when the Fourier transformation is carried out in dimensional units. Observe that there are substantial differences between the sub-networked coordinate space and the mathematically constructed hybrid space. Firstly the spaces obey different algebrae. Secondly hybrid space is an information space possessing matter; it has no events; its scale (in natural atomic units) is derived from the Fourier transformation. In contrast coordinate space is the coherent representation of the underlying network and this space is naturally based upon Planck units.

Hence when we compare the different spaces we do so using different systems of units that provide a non-linear change in scale. This non-linearity is illustrated by comparing time differences and the age of the universe in different unit systems. From equations (11.21) and (11.22) we find

$$\frac{\Delta t_{u_2}}{\Delta t_{u_1}} = \frac{3}{2}\frac{A_{t_{u_2}}}{A_{t_{u_1}}}. \qquad (17.76)$$

Here the numerical factor of $^3/_2$ comes from the order of the power in the equation relating $A_{t_{u_1}}$ and $A_{t_{u_2}}$ ($A_{t_{u_2}} \sim A_{t_{u_1}}^{3/2}$).

Considering a unit of information, the reciprocal transformation scales the complementary variables as $\hbar / m_u$. After a suitable unit conversion, the discrimination volume of the phase space is

$$x_{\min} w_{\max} = x_{\max} w_{\min} = \frac{\hbar}{m_u} \frac{{}_x\breve{u}_2}{{}_x\breve{u}_1} \qquad (17.77)$$

This result confirms that the atomic unit of distance, ${}_x\breve{u}_1$, must be $\hbar / m_u$. Hence atomic units display the time dependent as shown in table 11.1.

Next, we will consider the measure of these ranges. We live in an expanding universe. However, as the contravariant distance is expanded the covariant distance is contracted. Then the contracting minimum hybrid-space distance in atomic units becomes

$$\left(H\ {}_t\breve{u}_1\right)^2 cA_t \,. \qquad (17.78)$$

Then we convert from atomic units to Planck units (using equation 17.76)

$$8_x\,\breve{u}_2 = \frac{3}{2}\frac{{}_t\breve{u}_1}{{}_t\breve{u}_2}\left(H\ {}_t\breve{u}_1\right)^2 cA_t \,. \qquad (17.79)$$

Apparently we do not exactly obtain the Planck distance unit, the doubling up of events when we fold the positive and negative time of coordinate space together delivers a factor of eight. Hence we recover equation (11.18) from which we may deduce the value for m_u.

17.5 Hypothesis Revealed

Eventual reality's description of nature has been revealed. The universe at its most basic is the apeiron. Using terms that we can relate to experience, we have created a model description of the apeiron as a network of events in which qualities are derived from the local arrangements. However, it may be that there is no known language or logic that can completely describe the apeiron. The apeiron is the foundation storey of a grand multi-storey edifice. Each elevated storey extracts from the storey below, a larger-scale portrayal of reality displaying more and more order, regularity and regulation.

On the first-elevated storey we have the beginning of our recognizable universe: space filled with an aethereal field interpretable as a musing in a universal langue. Thus language, logic, mathematics and space are products of the first-elevated story of reality. Consequently this first-elevated storey and all subsequent storeys are naturally effable.

The second-elevated storey sees matter and conventional space forged from the aethereal field and foundation space. However, matter exists only as a constituent of the whole – a whole that seems to be less fascinating than the sum of its parts. Particulate matter is a constituent of a statistical description of a less-informed immaterial aether. There are interesting analogues of this many/one duality: life forms can be described as constituents of an ecosystem that is not alive; events have been seen to acquire relational qualities not possessed by the whole apeiron. Each constructed storey fashions it own intrinsic concepts to suit its appearance. The new conceptual clothes may cover and hide the old clothes or the new clothes may recycle the old clothes into a different garment. When the whole and its parts reside on different storeys of reality, then new concepts may emerge in the whole and old concepts may vanish from the whole. Thus it is unfair to ascribe a comparative measure of content to viewpoints on different storeys.

Likewise the aethereal field's constituent linguistic expressions hold a rich expression of meaning that is lacking in the aether as a whole. This langue provides for the mentality of the universe. However, mind requires more than a rich capability for mental expressions, mind requires an associated memory. Memories need to be sustainable and hence are laid down by quantum measurement in classical material records. This connection anchors mind to matter. Furthermore the connection discloses the distant prospect for building quantum machines with genuine intelligence!

Astoundingly in the quantitative scientific realm there is but one equation of everything (quantitative). This equation is the first principle of the quantitative universe; it is an updated form of Fermat's principle

$$\delta S = 0 . \qquad (17.80)$$

By making different approximations, we saw how this same equation can be used to derive both quantum mechanics and general relativity.

In totality the propositions of eventual reality are utterly simple; the predictions and explanations of eventual reality are inspiringly expansive. The outcomes of this work seem too spectacular to be possibly true. Could the work be flawed by errors? Undoubtedly I have overlooked many important issues. However, science often advances through making errors. It should not matter if there are some flaws; mature theories are rarely the work of a single individual. Eventual reality does deliver a substantial inventory of utterly unique explanations. The nature of fundamental reality has been obscured from science for so long a time that recently physicists have been challenging science's capacity to describe a fundamental reality. Eventual reality shows that: science can spawn a description of fundamental reality; a science-based study can explain why there are laws of science; science can explain the dimensionality of space; science can explain the symmetry of fundamental particles; science can explain the union of general relativity and quantum mechanics; science can explain the second law of thermodynamics; etc. However, the hypothesis of eventual reality goes much deeper explaining the existence of science, logic, mathematics, philosophy and ultimately the universe and the theory of eventual reality itself. Furthermore eventual reality explains the mind and explains the interconnection of mind and matter.

Notes

[1] Traditionally we treat *very much greater than* as meaning greater than ten. In entropy expressions, this superlative inequality also has to apply to the logarithm.

[2] Recall that a tensor representation can only describe matter classically. We need to generalize the mathematics to fully describe matter.

[3] We explicitly show the metric g^{ij} by writing c_i and ϕ_i as covariant variables then, for the second term, use integration by parts to switch the scope of the differential.

[4] These are the symmetries noted by Dirac and discussed in section 10.6.

[5] We choice the Ricci scalar of curvature to unit power since the Ricci scalar has the same coordinate power as the affinity.

[6] If distance units are time dependent then time and distance are not entirely independent.

Epilogue to Part 3

This is the scientific information age. Not surprisingly the scientific community generates a babble of ideas. Anyone who stops to listen and consider all relevant under-developed speculation will inevitably follow a random walk through magnificent gardens displaying exotic fruits. Whilst being enjoyable and providing exercise, this walk will not take the traveller far very quickly. Indeed I suggest that this is what has been happening in physics for some time. It would have been possible for the hypothesis of eventual reality to be developed at any time during the last third of a century (including by the author during this time). Of course social changes have not helped: pressures of publishing accountability and the shrinking world have complicated our personal worlds and served to blunt the traditional belief that nature is simple. Hence in our random walk (over the last three decades) we have encountered p-brane theories satisfying as yet unimagined mathematics, hierarchies of non-encounterable multiverses, cosmological missing links from cosmic inflation to dark matter and dark energy, intriguing anthropic artistry, etc.

In contrast, the current work has maintained a disciplined focus and has kept the journey's heading in a consistent direction. In this manner, we have completed the intended journey and located a hypothetical font of all knowledge - the theory of eventual reality. The theory is simple, elegant and radical.

It might seem that every individual feature of eventual reality's worldview has been proposed before. However, what distinguishes eventual reality from the clamour of suggestions about what constitutes reality is the theory's essential non-quantitative foundation and its expansive and consistent worldview. In addition the theory emphasizes and formalizes the hierarchical structure of explanations (the multistorey edifice of reality).

This still-juvenile hypothesis has delivered a parameter-free and self-justifying complete theory of reality. Furthermore, eventual reality has many other remarkable achievements. Eventual reality pulls together all of the sciences, philosophy, mathematics, logic and even linguistics into a unified whole framework. In philosophy, we have healed the rift between empirical and rational arguments. We have provided answers to many of the fundamental questions of existence. We have also dismissed the conflict between our physical worldviews and philosophy. In logic, we have identified apposite logical expressions upon which logic and mathematics are built.

In Physics, eventual reality has fixed the fractures existing within quantum mechanics and also between the paradigms of quantum mechanics and relativity. Indeed eventual reality has subsumed both quantum mechanics and general relativity into one theory of everything which is expressible as a single equation. Eventual reality has revealed the symmetries of the fundamental particles and explained their quantum numbers. Eventual reality has provided a simple expression for the cosmological constant, the expansion of the universe and variations of universal constants. However, most notably eventual reality requires that the universe to display mind, matter and space.

Now, there is no unique way of doing theoretical physics or indeed any discipline of science. Like every human activity, every strategy has its strengths and associated weaknesses. Hence a mixed approach invariably offers the best solution. However, there are attractive fashions that promote commonality and act against diversity. From the perspective of the mind, there are typically four styles of approaches (to studying fundamental questions) that will most likely appeal differently to scientists with differing personalities. Of course no one is entirely motivated by a single style of working.

One approach appeals to the emotions. We are all enthralled by a beautiful and elegant theory that just has to be correct. This approach seeks out the elegant and powerful. An example of an elegant and powerful equation is provided by Laplace's equation which was once fleetingly considered as a solution of the universe. Laplace's equation describes the generic form for a scalar field in a source-free region of 3-space:

$$\nabla^2\Phi = \frac{\partial^2\Phi}{\partial x^2} + \frac{\partial^2\Phi}{\partial y^2} + \frac{\partial^2\Phi}{\partial z^2} = 0 \cdot$$

For instance, Laplace's equation describes the Newtonian $\sim 1/r^2$ weakening of the gravitational force. In contrast, we have shown that Fermat's elegant theorem has an all embracing scope and it is eminently useful and powerful. Indeed what could be simpler than a qualitative world based upon

$$\delta S = 0 \cdot$$

Parmenides should celebrate! The structuring of the universe is changeless.

Another approach appeals to the logical analytical mind. This approach produces the crisp precision of formal mathematics and inescapable logic. This is exemplified by the work of many mathematicians who have moved into fundamental areas of theoretical physics. Well, eventual reality has a firm logical if not mathematical foundation. Furthermore the theory also proposes a firm foundation for mathematics.

A third approach demands rigid acceptance of an empirically founded necessity. Our beliefs are very powerful and they colour our view of the world about us. This approach is typified by the dogmatic insistence that only things that we can see are actually real, an insistence that historically railed against acceptance of the atomic hypothesis. Uniquely eventual reality lays claim to all of empirical reality: mind, matter and space.

The fourth approach perhaps best describes the style of the current work. This fourth approach focuses upon the creation of a new deeper picture providing a wider understanding of the issues. This more philosophical approach involves reconstructing existing knowledge into a simpler and differently assembled framework of ideas – the big picture.

This suggested list of scientific drivers reveals the potential weakness of the current hypothesis. The presented wide-ranging picture of reality is this document's strength and weakness; the detailed expositions are derivative to the global explanations. The current document is the first introduction to the hypothesis of eventual reality, much has only been sketched. It would be quite remarkable if many particulars of the expansion of this picture to describe aspects of reality were found not to be in need of some substantial revision. Undoubtedly this work does contain many errors and omissions. Nevertheless I commend it to you for your further consideration.

Finale

When undertaking a work such as this, the question arises as to when the author should stop developing the theory and publish the initial results. Now, acceptance criteria for scientific works vary according to the reasonableness and the exceptionality of the claim. Throughout this work I have argued strongly that the claims of the theory are very reasonable. Unfortunately the theory appears to be exceedingly exceptional and consequentially the theory should fall prey to the familiar maxim that *exceptional claims require an exceptional degree of evidence or proof*. Thus I have felt it necessary not to rush into print and to give a reasonably comprehensive account of the theory so that the theory's truly exceptional capabilities can be clearly observed. However, the result of this work is a theory that complements existing science expanding the horizons of knowledge far beyond the quantitative world. This complementarity makes the claims of the theory less exceptional and hopefully in the absence of any competition quite acceptable. Hence although there is never a perfect time to publish, the author feels that the theory has reached an acceptable level of maturity and this presentation is as clear as I can make it.

Science is a series of questions. Oftentimes there are answers but the answers always seem to raise even more questions. Indeed we have emphasized this point by compiling several lists of questions and answers, viz. the Copenhagen interpretation of quantum mechanics (section 1.2.2), the slippery questions (epilogue part 1), mind-body problem, core of quantum mechanics (section 10.10.1), the meaning of reality (section 13.4), basis of physical reality (section 14.2), how we view fundamental questions (epilogue to part 3), not to mention *the* ultimate and penultimate questions with which we started this document. In this discussion, I am leading up to perhaps my most contentious and final compendium of questions.

Everyone is entitled to believe what they like to believe. However, the various strands of our societies are manipulated by self-interested groups of power wielders who believe in their own benevolence and seek to promulgate beliefs that enhance their power and status. When Galileo defiantly murmured under his breath 'but it (the Earth) still moves', he was asserting a truth of science; Galileo was not undermining religion but he was refuting an insecure fallible authority that insincerely claimed infallibility. Scientific knowledge has grown since Galileo's time, making science an even more attractive target for short-sighted political, social, economic and religious interests. Scientists are humble and accepting, but science is not. Politicians believe in strategic compromise; science accepts only an uncompromising impartial truth. The long-standing orchestrated attacks on evolution in the United States (eg the current *Intelligent Design* 'debate') represent futile attempts by politicized groups to use societal power primarily to appropriate some of the credibility earnt by science but secondarily to tarnish the status of science in society. Perhaps these attempts may now bear the epitaph *too little too late* – for the hypothesis of eventual reality is here.

Eventual reality is a theory of natural philosophy and natural philosophy fully encompasses science. This document has been primarily about science and the interpretation of science. However, as these remarks demonstrate, the document also strays into more dangerous territories. For instance, eventual reality can say much about what is a tenable rational belief. Therefore at the risk of provoking greater enmity from distressed souls, let me discuss in my own fallible way whether the hypothesis of eventual reality is compatible with the hypothesis of a supreme supernatural being. The reader may choose either, neither or both hypotheses. We shall accomplish this compatibility task by posing a set of questions that explore the potential properties of a deity. I will proffer opinions on these questions revealing where there is scope for the spiritual.

1) *Is there an omnific creator? Is there an initiator, a first cause of everything in the universe?*
Eventual reality does not recognize anything other than the universe. Thus there is nothing external that can bring the universe into being. Fortunately nothing is required, since existence of the universe is logically required, as tautologically it seems is rational logic. However, one can adopt another perspective in which the (quantitative) universe is identified with coordinate space. Then the apeiron, existing without space, summons into being the quantitative universe. In this sense, the apeiron is omnific.

2) *Is there an omnipotent being? Is there an all-powerful influence?*
This question is like the old chestnut: what happens when an irresistible force meets an immoveable object? If an omnipotent being cannot create something that he cannot change, then he cannot be omnipotent. However, if he can create it, then he cannot be omnipotent either. Eventual reality validates the use of rational logic. Hence the question answers itself almost independently of eventual reality. However, from the perspective of the apeiron nothing is impossible. The apeiron side steps logical imperatives which first arise in the apeironic offspring known as coordinate space. There can be no all-powerful influence in coordinate space; for instance, the dimensions of coordinate space are fixed. Nevertheless there is one restriction placed upon the apeiron – the apeiron must exist.

3) *Is there a setter who decides the immutable laws of nature? Is there a designer of the universe?*
Well, this designer would be part (or whole) of the universe. However, eventual reality offers no options for the immutable properties of the universe. Coherency fixes the only possibility for these non-contingent universal properties and laws. Coherency has no choice.

4) *Is there a fateful purpose for which everything exists? Is there a final outcome which stands as a reason for everything?*
While eventual reality does allow the temporal aspects of causality to be turned around, the beginnings and endings cannot be held responsible for what happens in between. Indeterminism and choice are indelibly embedded in quantum-level reality.

5) *Is there a widespread untameable will that is free of external dictates? Is there an all-embracing independent influence that can direct but not be directed?*
At the fundamental level, everything in the universe is interconnected either directly or indirectly. This reality is so intertwined that, in principle, changing anything changes everything. An influence that can direct an action will receive a feedback influence (even if that influence is relatively small). There can be no complete independence, no truly independent influence on the happenings of the universe.

6) *Is there a builder of the universe? Is there a decider of contingency?*
Let me give a specific example to explain the question: Are the mass of the Earth and our planet's place in the galaxy contingent accidents or are they the considered opinion of a great builder who is working within the constraints of the laws of nature? Our answer cannot be dismissive since everything in the universe is not corporeal. Further, the temporal evolution of the statistics of the aethereal field is indeterminate and our awareness is sourced from just such indeterminacy. Clearly as mixing occurs across the branches of space, the universe certainly delivers decisions (random or otherwise, intentional or otherwise). Indeed all of our physically observable reality is interconnected through the window of the big bang. Thus the aether can be construed as building our universe.

7) *Is there an omniscient entity? Is there a complete record of all proceedings and all knowledge?*
The aether is the clay from which our appreciations of mind and matter are crafted. The aethereal field spanning the branches of hybrid space certainly holds all knowledge of all occurrences, of all matter, of all thoughts, insights, feelings, judgements, opinions, understandings, etc. Certainly the aether is omniscient.

8) *Is there an omnipresent shared awareness throughout space? Is there an insubstantial conscious element unifying the natural universe?*
The statistics of an unwilling aethereal field may generate matter and its dynamics. However, the aether has much greater capability. The consistent application of a will requires a memory and methods of access and processing. However, these are already features of the aether.

9) *Is there a supreme judge presiding over all of life's actions? Is there an ultimate voyeur assessing all of our decisions?*

The same psychological dimensions are available to any universal super mind as are available to us. Further, our psyches have to be part of this super mind. Nonetheless if we are being judged by an independent superior mind sharing the same aether, we should be able to feel, hear or participate in the judgement.

These discussions of the nine questions, underscore how eventual reality only claims jurisdiction over all rational and all empirical knowledge. Alongside these claims there is room for further (Polonius-style) interpretations founded upon faith.

Now, yesteryear the Earth was at the centre of the creation. Yestermonth Euclidean space was absolute. Yesterday matter and energy were fundamental. Today we must consider eventual reality. Who knows what tomorrow may bring.

Science is a valiant quest to acquire an understanding of the universe, about finding our place in the great scheme of things. Through eventual reality we find that we do belong. Finally science can comprehensively challenge the world of misconceptions and illusions. Let the empire of science strike back to rid the world of ignorance and deceit. The (old) paradigm is dead; long live the (new) paradigm.

References

Adler, I (1958) *The New Mathematics* (New York: Signet Science) 192pp

Bartusiak, M (2004) *Origin & Fate of the Universe* Special Edition of *Astronomy* pp 064-071

Bronowski, J (1973) *The Ascent of Man* (London: BBC) 448pp

Devlin, K (1988) *Mathematics: The New Golden Age* 2nd Edition (London: Penguin Books) pp 73-74

Dicke, R H (1961) *Nature* **192** 440-441

Dirac, P A M (1973) *Die Naturwissenchaften* **60** 529-31, reproduced in Landsberg (1982) pp 109-113

Greene, B (1999) *The Elegant Universe: Superstrings, Hidden Dimensions, and the Quest for the Ultimate Theory* (London: Vintage) pp 283-319

Heisenberg, W (1958) *Physics and Philosophy: The Revolution in Modern Science* (Great Minds Series 1999) (New York: Prometheus Books) pp 44-58

Iyanaga, S and Kawada, Y (1977) Editors, *Encyclopedic Dictionary of Mathematics* English Translation (Cambridge Massachusetts: MIT Press) 2 Vol set

Kaku, M and Thompson, J (1997) *Beyond Einstein: The Cosmic Quest for the Theory of the Universe* (Oxford: Oxford University Press) 230pp

Landsberg, P T (1982) *The Enigma of Time* (Bristol: Adam Hilger) 248pp

Leggett, A J (1977) in *The Encyclopedia of Ignorance* Ed R Duncan & M Weston-Smith pp 101-9 reproduced in Landsberg (1982) pp 149-156

Penrose, R (2004) *The Road to Reality: A Complete Guide to the Laws of the Universe* (London: Jonathan Cape) 1094pp

Percus, J K & Yevick, G J (1958) *Phys Rev* **110** 1-13

Popper, K (1974) *Unended Quest: An Intellectual Autobiography* (Glasgow: Collins) pp 180-187

Rees, M (1997) *Before the Beginning* (London: Free Press) 282pp

Russell, B (2004) *History of Western Philosophy* (London: Routledge) 778pp

Schrödinger, E (1950) *Space-Time Structure* (Cambridge: Cambridge University Press) 119pp

Shand, J (1994) *Philosophy and Philosophers: An Introduction to Western Philosophy* (London: Penguin Books) 348pp

Sivaram, C (1982) *Am. J. Phys.* **50**(3) 279

Smolin, L (1997) *The Life of the Cosmos* (London: Phoenix) 440pp

Sorensen, R (2005) *A Brief History of the Paradox: Philosophy and the Labyrinths of the Mind* (New York, Oxford University Press) 394pp

Stokes, P (2002) *Philosophy ~ 100 Essential Thinkers* (London: Arcturus Publishing Ltd) 208pp

Weyl, H (1952) *Space-Time-Matter* 4th edition English translation by H L Brose (New York: Dover Publications) 330pp

Index

D

E

F

G

H

I

N

O

P

Q

R

S

T

U

V

W

Z